Lymphocyte Immunization Therapy (LIT) in Reproductive Failures

Mohan Raut • Mugdha Raut

Lymphocyte Immunization Therapy (LIT) in Reproductive Failures

New Horizons

Springer

Mohan Raut
Dr. Raut's Centre for Reproductive Immunology
Mumbai, India

Mugdha Raut
Dr. Raut's Centre for Reproductive Immunology
Mumbai, India

ISBN 978-981-19-2959-5 ISBN 978-981-19-2960-1 (eBook)
https://doi.org/10.1007/978-981-19-2960-1

This Springer imprint is published by the registered company Springer Nature Singapore Pte Ltd.
The registered company address is: 152 Beach Road, #21-01/04 Gateway East, Singapore 189721, Singapore

This book is dedicated to our teacher, mentor, friend, philosopher, and guide Dr. Usha Krishna. We would not be where we are today, in the field of reproductive immunology, without her.

Preface

This is the first-ever book to be written on alloimmune aspects in "unexplained" repeated pregnancy loss (RPL), recurrent implantation failure (RIF), and the role of lymphocyte immunization therapy (LIT) in these conditions.

Pregnancy is a unique condition where the embryo/fetus grows in an immunologically different environment of the mother without any immune reaction against it. The fetus actually is a "semiallograft," meaning it enjoys special status in the mother's body. This "paradox of pregnancy" has intrigued biologists, immunologists, reproductive immunologists, physicians, and obstetricians for decades. And over the years, many theories have been proposed to explain this complex phenomenon. Clinical research is now being combined with ongoing research in reproductive immunology to understand the mechanism behind successful pregnancies. This is important because now it has been established that in couples with unexplained RPL/RIF, alloimmune incompatibility plays an important causative role. It is found that in more than 50% of RPL and RIF cases, the possible reason could be the alloimmune rejection of pregnancy. Numerous studies have also been done to examine the impact of different immunomodulatory treatments on the management of RPL/RIF.

LIT is the first and most popular immunomodulation modality. It has been extensively researched and recommended all around the globe, and multiple studies have demonstrated the effectiveness. In fact, it was also one of the first treatments recommended for uRPL/uRIF.

A big reason to write this book was the lack of awareness about this excellent method of treatment called LIT. This book aims to inform about how LIT was developed, the mechanisms through which LIT acts, and the procedure to perform LIT. Since LIT has been surrounded by different controversies from time to time, there are important chapters in this book which clear the air around them and talks about the latest meta-analyses that prove the efficacy of LIT.

We have also shared reviews of various studies related to LIT, conducted over the last 20 years, which will help readers understand LIT better. The book also addresses the most important dilemma to clinicians, which is the diagnosis of alloimmune factors. We have talked about the various tests that are offered for the diagnosis. Our association with LIT started in 1991, when Dr. Mohan was a part of the multicentric double-blind study on the role of LIT in RPL, conducted by Nuffield Department of

Obstetrics (Prof. Redman), Oxford. Now, we are doing LITs at our center—Dr. Raut's Centre for Reproductive Immunology, in Mumbai, India, with a patented process.

We hope that this book will help readers in understanding LIT better and clear the doubts and uncertainties associated with the procedure. We truly believe that research is an integral part of science, and it contributes greatly to the progress of science. Reproductive immunology is an ever-expanding field, and we will witness more research in this field in the future. It is very important to convert this research into clinical practice so that it can help mankind. We are also sure that LIT will also see many upgrades with continuous research, and it will help an increasing number of couples who are suffering from the unfortunate problem of "unexplained" RPL/RIF, and they will be able to fulfill their dream of having their baby.

As the process of LIT is under patent protection, all the details could not be revealed in the book, but all procedures and facts are discussed, along with their safety regulations. Towards the end, there are case studies of the LIT process and our thoughts on the future of LIT.

Mumbai, India Mohan Raut
Mumbai, India Mugdha Raut

Acknowledgments

It is our privilege to announce that *Lymphocyte Immunization Therapy (LIT) in Reproductive Failures: New Horizons* is the first book exclusively dedicated to the subject of Lymphocyte Immunization Therapy (LIT). We have received help, support, and encouragement from a lot of people. We appreciate and acknowledge all of them from the bottom of our hearts.

However, there are some who deserve a special mention, and our expression of gratitude will not be complete without that.

We would like to extend our deepest gratitude to Dr. Usha Krishna, because of whom we were introduced to the field of Reproductive Immunology. She has guided us throughout our journey in this field and in setting up of our Reproductive Immunology center.

We would like to express our deepest thanks to Dr. O. B. Christiansen, ex-secretary and currently Councilor, European Society for Reproductive Immunology. He had inaugurated our Reproductive Immunology center in Mumbai and that paved the way for our work related to LIT, which has now culminated into this book.

We would like to thank Dr. Joanne Kwak-Kim, past president, American Society for Reproductive Immunology, for her constant encouragement in our work related to LIT.

The concept of this book came into existence due to the suggestions of Dr. Abhinav Shrestha Jha, ex-associate editor, Springer Nature, and an immunologist. We would like to pay our special regards to him for that.

This book would not have been completed without the help of our daughters, Adv. Nupur, who practices law in Delhi, and Natasha, molecular biology and data science student from UC Berkeley. We would like to recognize their invaluable assistance in this venture with a special thanks to them.

We are very grateful to Mr. Piyush Bhatia and his team for their timely help.

We are also thankful to the Springer Nature team, especially Dr. Bhavik Sawhney (Associate editor—biomedicine) and Mr. Kumar Athiappan (Project Coordinator—Books), for their help with a lot of patience!

In the end, we would like to express our gratitude to our teachers, parents, patients, well-wishers, and the Almighty!

Contents

Part II Lymphocyte Immunization Therapy

About the Authors

Mohan Raut, MD, DGO, FCRI (ASRI), received his graduation and post-graduation from the premier institute Seth G.S. Medical College and KEM Hospital, Mumbai.

He has been in private practice in obstetrics and gynecology for the past 33 years, with special interest in the field of reproductive immunology, especially related to repeated pregnancy loss, recurrent implantation failure, and unexplained infertility.

In 1992, Dr. Mohan Raut was a part of the International Multicentric Research Project in association with Oxford University on the role of lymphocyte immunization therapy (LIT) in recurrent miscarriages.

Later, he had been involved in the LIT program at Sir H.N. Hospital, Mumbai, under the guidance of Dr. Usha Krishna.

He has contributed a chapter on Immunotherapy in Repeated Miscarriages in the book *Pregnancy at Risk: Current Concepts* published by FOGSI.

He has started India's first reproductive immunology center, Dr. Raut's Centre for Reproductive Immunology (previously known as ICPRM), with Dr. Mugdha Raut and uses a patented process for LIT procedure.

He has published papers and has spoken in National and International conferences on the topic *Immunomodulation in Repeated Miscarriages and Recurrent Implantation Failures*.

He has been honored with many awards for his work in the field of reproductive immunology, including the Prime Time Global Healthcare Excellence Award for work in the field of immunotherapy (2017).

The American Society of Reproductive Immunology has awarded him Fellowship in Clinical Reproductive Immunology (FCRI) for his work in the field of clinical reproductive immunology.

He is the Guest co-editor of the American Journal of Reproductive Immunology (AJRI) and a member of the editorial board of Journal of Evidence Based Women's Health Journal (EBWHJ), Egypt.

He is the managing trustee of Dr. Raut's Immunotherapy Foundation (a registered charitable trust) and arranges programs to create awareness about reproductive immunology and provides affordable medical care.

Mugdha Raut is MD, DGO, FCRI (ASRI), received her graduation and post-graduation from the premier institute Seth G.S. Medical College and KEM Hospital, Mumbai. She was a gold-medalist in her graduation program.

She has been in private practice in obstetrics and gynecology for the past 30 years, with special interest in the field of reproductive immunology, especially related to repeated pregnancy loss, recurrent implantation failure, and unexplained infertility.

Dr. Mugdha was trained in Infertility under Dr. Mehroo Hansotia at Nowrosjee Wadia Maternity Hospital, Mumbai. She was associated with Dr. Usha Krishna at the KEM Hospital for her work in the field of repeated miscarriages and BOH (Bad Obstetric History).

Her passion and vision in the field of reproductive immunology resulted in the establishment of Dr. Raut's Centre for Reproductive Immunology (previously known as ICPRM), India's first center in reproductive immunology. The center uses a patented process for LIT procedure.

She has published papers and has spoken in National and International conferences on the topic *Immunomodulation in Repeated Miscarriages and Recurrent Implantation Failures*.

She has been honored with many awards for her work in the field of reproductive immunology, including the Times of India Healthcare Achievers Award for the innovative work in the field of reproductive immunology (2018).

The American Society of Reproductive Immunology has awarded her Fellowship in Clinical Reproductive Immunology (FCRI) for her work in the field of clinical reproductive immunology.

She is a member of the Education Committee of the American Society of Reproductive Immunology (ASRI) and also a member of the editorial board of Journal of Evidence Based Women's Health Journal (EBWHJ), Egypt.

Dr. Mugdha was awarded by the prestigious Femina magazine as the TOP 10 WOMEN WHO MATTER and was recognized by the prestigious Cosmopolitan magazine as among the TOP 10 INFLUENTIAL WOMEN (March 2014).

She is the trustee of Dr. Raut's Immunotherapy Foundation (a registered charitable trust) who arranges programs to create awareness about reproductive immunology and provides affordable medical care.

Part I

Reproductive Immunology and Reproductive Failure (Repeated Pregnancy Loss/Recurrent Implantation Failure/Unexplained Infertility)

Introduction to Reproductive Immunology and Reproductive Failure

1

Abstract

Infertility, recurrent implantation failures (RIF), and recurrent pregnancy loss (RPL) have similar causative factors, diagnostic tests, and treatment protocols to a larger extent. This is more evident when these conditions are "unexplained." As the role of alloimmune factor is significant in these conditions, unexplained RPL and unexplained RIF are considered together. This part discusses reproductive immunology and diagnosis of alloimmune factors by immunological tests. The role of lymphocyte immunization therapy (LIT) in these conditions is discussed in the next part.

Keywords

RPL · RIF · Alloimmune factor · Unexplained reproductive failures · LIT

Infertility and RPL (recurrent pregnancy loss) are two separate entities that are typically managed in different ways. However, there are similarities between the two conditions. The underlying causative factors are common to a considerable extent in both these conditions. The diagnostic and treatment protocols are also similar for a considerable number of people in these two groups [1]. This becomes more evident in presence of unexplained RPL and unexplained infertility, especially unexplained recurrent implantation failures (RIF). The role of the alloimmune factor is important in both these cases of "unexplained" reproductive failure. Hence, "unexplained" RPL and "unexplained" RIF are considered together in this book, with respect to possible underlying mechanisms, diagnosis, and treatment. The role of Lymphocyte Immunization Therapy in these conditions is also discussed in detail.

Lymphocyte Immunization Therapy, also known as LIT, involves injecting white blood cells from the prospective father or random donor(s) into the skin of the prospective mother in order to prepare her immune system for pregnancy. LIT aids the mother's immune system in developing immunological tolerance toward

M. Raut, M. Raut, *Lymphocyte Immunization Therapy (LIT) in Reproductive Failures*, https://doi.org/10.1007/978-981-19-2960-1_1

genetically foreign tissues. While there is some controversy surrounding lymphocyte immunization therapy (LIT), it offers hope for women suffering from unexplained recurrent miscarriage or recurrent IVF failure.

We will start with a brief discussion about RPL and RIF and then move to discuss the unique features of the immune system. This will be followed by an interesting insight into the immunology of pregnancy and how immunological balance helps to protect it. The role of different immunological tests in selecting couples for immunomodulatory treatments and the diagnosis of alloimmune factors will also be discussed in broad detail.

Reference

1. Kumar D, Harlev A. Recurrent pregnancy loss and infertility: a time for change. Austin J Obstet Gynecol. 2018;5(1):1090.

2 Recurrent Pregnancy Loss and Recurrent Implantation Failure

Abstract

Recurrent Pregnancy Loss (RPL) affects 2–5% of the population, and 10% of all the patients undergoing IVF treatment end up with Recurrent Implantation Failure. Nearly 50% of these patients are not aware of the cause of their pregnancy loss and so their cases are often referred to as "unexplained." This chapter discusses the definitions, incidence, types, and classifications of RPL and RIF. It also talks about the importance of biochemical pregnancy and the concept of RPL attributable to chance.

Keywords

Recurrent pregnancy loss · RPL · Recurrent implantation failure · RIF · Unexplained infertility · Definition of RPL · Definition of RIF · Incidence of RPL/RIF · Types of RPL · Biochemical pregnancy · RPL by chance

2.1 Recurrent Pregnancy Loss

2.1.1 Introduction

Recurrent Pregnancy Loss (RPL) can be a heterogeneous condition. This affects women of all ages and with a variety of conditions. Because the cause of RPL is not always known, it is one of the most frustrating and challenging conditions in reproductive medicine. Also, the evidence for diagnostic and therapeutic methods is not always reliable. Many studies that examine the causes, management, and evaluation of RPL are flawed [1]. That is why, RPL treatment can often be frustrating.

M. Raut, M. Raut, *Lymphocyte Immunization Therapy (LIT) in Reproductive Failures*, https://doi.org/10.1007/978-981-19-2960-1_2

2.1.2 Definitions

The heterogeneity and ambiguity of recurrent pregnancy loss (RPL) begins with the definition of RPL. For the purpose of appropriate evaluation of RPL, the pregnancy is classified as clinical pregnancy recorded by ultrasound, or histopathological examination [2].

The number of pregnancy losses is the accepted criterion in labeling a case as RPL. American Society of Reproductive Medicine (ASRM) says that there have to be two or more than two miscarriages to be RPL.

The European Society of Human Reproduction and Embryology, however, says that three or more miscarriages should be considered as RPL [3].

With the ASRM definition, the incidence of RPL is 2–5%, while with ESHRE definition RPL affects 1–2% of pregnant women [4, 5].

RPL-prone women may have pregnancies that resulted in live births. Based on these factors, there are three types of RPL (see Table 2.1).

Primary RPL: Women who have lost two or more pregnancies in a row, but no further than 20 weeks.

Secondary RPL: Women who have lost two to three or more pregnancies after a pregnancy past 20 weeks of gestation that may have resulted in stillbirth or live birth.

Tertiary RPL: Women who have suffered multiple pregnancy losses prior to a pregnancy that was past 20 weeks of gestation, after which there were at least two or three more pregnancy losses [6].

2.1.3 Biochemical Pregnancy and RPL

There has been much debate about whether biochemical pregnancy loss must be included in the definition of RPL or not. The ASRM (American Society of Reproductive Medicine) suggests that pregnancies that can be recorded through ultrasound or histopathology only should be considered as RPL [7].

In the overall population, the claimed percentage of pregnancies that end in preclinical loss is anywhere between 60 and 80%. Biochemical pregnancies are very common. A second study found that nearly 20% of women may experience three biochemical pregnancies due to chance alone and they do not require to be investigated [8].

Despite all this, many clinicians include biochemical pregnancies in the RPL definition because women are anxious to undergo proper evaluation for RPL, and

Table 2.1 Types of RPL

Type	Pregnancy losses	Pregnancy beyond 20 weeks (Live birth/still birth)
Primary	≥2or 3	Nil
Secondary	≥2 or 3	Yes—Before Pregnancy Losses
Tertiary	≥2 or 3	Yes—In between Pregnancy losses

they rely on nonprescription sensitive urine pregnancy tests to confirm biochemical pregnancy.

A study from the European Society for Human Reproduction and Embryology (ESHRE), which examined RPL patients, revealed that any early loss that was confirmed by a positive Beta hCG testing had a negative prognostic significance equal to that of a medical miscarriage. It revealed that biochemical pregnancies make up about 37% of recorded pregnancies, and that each additional non-visualized pregnancy decreases the chance to have a live baby by at most 10% [10]. This supports the idea of biochemical pregnancies being included in the RPL diagnosis.

2.1.4 Diagnostic Criteria for RPL Definition

Different RPL clinics worldwide mention different diagnostic criteria in their national guidelines [9–12].

The criteria are as follows:

More than or equal to three consecutive pregnancy losses before 24 weeks.
More than or equal to two consecutive pregnancy losses.
More than or equal to three consecutive miscarriages and biochemical pregnancies.
More than or equal to three not necessarily consecutive pregnancy losses before 24 weeks of pregnancy.

2.1.5 RPL Attributable to Chance

A group of women with RPL will always include some women who continue to have a pregnancy, resulting in a live birth. These women can be put in a group where RPL is due to pure chance [13, 14]. Even after all investigations are done, almost 50% of women suffer from unexplained RPL. It is important to know the difference between RPL due to chance and RPL that is genuine and unexplained. The latter will need to be investigated and can be treated effectively.

It is interesting to see how the probability of an unexplained RPL rises with increasing numbers of RPL-related miscarriages. This can be calculated by a simple formula [13].

Spontaneous miscarriage loss rate = m
Number of pregnancies is n
Then,
Recurrent Pregnancy Loss rate = m^n

So, as seen in Table 2.2, as the number of miscarriages increases, the percentage of women suffering RPL due to chance goes on decreasing (see Table 2.2) [13, 15]. These women may have etiological issues that are not detectable by the protocols we have today. These women are often younger, and they have more

Table 2.2 Incidence of RPL occurring by chance for women with 1, 2, and 3 miscarriages

Recurrent pregnancy loss rates (occurring by chance) %			
Age (year)	1 Miscarriage	2 Miscarriages	3 Miscarriages
20–24	11	1.21	0.13
25–29	12	1.44	0.17
30–34	15	2.25	0.34
35–39	25	6.25	1.56

Data by Saravelos SH, Li TC [13]; and Nybo Andersen AM, Wohlfart J, Christens P, et al. [15]

Table 2.3 Characteristics of women with type I and type II unexplained RPL

Pregnancy losses				
Age		Number	Types	Karyotype
Type I	Older ≥30	<3	Includes biochemical	Abnormal Karyotype of conceptus
Type II	Younger ≤30	>3	Clinical only (FH+)	Remains unidentifiable (N)

Data from Saravelos SH, Li TC [13]

miscarriages. Hence, as the number of miscarriages increases, there is more chance of miscarriage in subsequent pregnancy. It is in these women of RPL that the role of the immunological system in the form of alloimmune interactions becomes important.

2.1.6 Types of RPL by Etiology

1. RPL of known etiology
   In RPL of known etiology, a known cause is found after doing the investigations, and they are then treated accordingly.
2. Unexplained RPL
   Unexplained RPL affects 50% of RPL women [16, 17].

2.1.7 Types of Unexplained RPL (See Table 2.3) [13]

1. Type 1 unexplained RPL:
   These are seen in older women with less than three miscarriages (including biochemical pregnancies and the likely pathology is abnormal karyotype of the fetus.)
2. Type 2 unexplained RPL:
   These are seen in younger women with more than three clinical pregnancies where the cause remains unidentifiable. This is the type of RPL in which miscarriage is most likely due to failure of alloimmune recognition and rejection

of pregnancy. Immunotherapy can play a very important role in the treatment of women who go through Type 2 unexplained RPL.

2.1.8 Incidence

It is essential to be aware of the chances of miscarriage, and the potential for miscarriages in the future. These are the main concerns for RPL couples. The overall risk of premature pregnancy loss in a first-time pregnancy is 15% [18, 19]. Due to the high rate of preclinical pregnancy losses, however, the real risk of early loss of pregnancy is between 50 and 60%.

Due to meiotic nondisjunction mistakes and chromosomal anomalies, the rate increases with age. The rate of early pregnancy loss at 20–30 years is 25% and almost doubles at 40 years [18, 19]. The predictive risk of miscarriage in a future pregnancy is about 14% after one miscarriage [20]. The risk of a pregnancy loss after two miscarriages is 24–29%, and after three miscarriages it is 31–33% [18, 21], up to as high as 48% [10]. As the number of miscarriages increases further, the chances of miscarriage in subsequent pregnancy increase rapidly. After four miscarriages it climbs up to 59%, and after five and more miscarriages it can be up to 77% [22, 23] (see Table 2.4). The negative prognostic effect of the number of previous miscarriages can be attributed to increasing maternal age and age-related risk factors, but even after correcting these variables, the number of previous miscarriages does adversely affect the risk of miscarriage in subsequent pregnancy and remains the strongest prognostic parameter [23–25].

2.2 Recurrent Implantation Failure

2.2.1 Introduction

The complex process of implantation of an embryo requires a synchronous development and interaction between the endometrium and the hatched blastocyst. If a gestational sac can be seen on ultrasonography, clinically successful implantation is considered to have occurred.

Table 2.4 Chances of miscarriage in subsequent pregnancy after previous miscarriages

No. of miscarriages	Chance of miscarriage in subsequent pregnancy (%)
0	11
1	14
2	24–29
3	31–48
4	59
≥5	77

Implantation failure refers to, when there is no evidence of implantation or hCG cannot be detected, or when proof of the implantation is present, hCG is detectable, but there is no gestational sac seen on ultrasonography [26]. It has been reported that in healthy fertile women spontaneous pregnancy is achieved in only about 25–40% of cases in the first cycle when the pregnancy is planned [27, 28]. Hence, implantation failure becomes important mainly in the context of assisted reproductive technique, especially in IVF or ICSI. In ART, the maximum implantation rate is between 40 and 60% [29]. Ten percent of couples undergoing IVF suffer from Recurrent IVF failure [30].

2.2.2 Definitions

RIF is difficult to define precisely because various centers use various criteria.

Here are the criteria used when RIF is defined:

Number of embryos transferred.
Number of ET procedures performed.

Accordingly, there are various definitions of RIF as follows:

1. Patients who have had more than 12 embryos transferred in different procedures, without becoming pregnant (Coulam, 1995) [31].
2. Patients with five to six unsuccessful cycles, including frozen embryo transfer, FET. There are approximately 10–15 embryos being transferred [UK].
3. Failure to achieve pregnancy after more than three failed transfers of embryos or transfers of 10 or more embryos (Tan 2005) [32].
4. Failure to become pregnant after more than three unsuccessful embryo transfers of high-quality embryos, or transfer of 10 or more embryos, in multiple transfers (PGD Consortium ESHRE) [33].
5. Failure to become pregnant after transfer of eight or more embryos (eight cell), or five or more blastocyst transfers [29].
6. Failure to achieve pregnancy after three single embryo transfers with good quality embryo (2006) [34].
7. Failure to achieve pregnancy after three or more ET or four embryos in women less than 40 years of age (2014) [26].
8. Recommendation of Istanbul Consensus: Failure of implantation of good quality embryos. Evaluation of good quality embryos should include morphokinetic assessment and ploidy status [35].

2.2.3 RIF Versus Recurrent IVF Failure

RIF is different from recurrent IVF failure. RIF is not only associated with unexplained infertility, but is also seen in the presence of known factors of infertility, like

Table 2.5 Types

Types	Cause
Multifactorial	Known causes (other than endometrium)
Endometrial	Thin endometrium (6 mm) with impaired vascularity
Idiopathic	Immunological (Allo-immune)

endometriosis, tubal factor, male factor, etc. Hence, recurrent implantation failure forms a subgroup of recurrent IVF failures [36].

Types of RIF [37] (see Table 2.5):

(A) Multifactorial RIF
Known causes are present. Anatomic defects in the reproductive system, male factor, genetic abnormalities in embryo, endocrine disorders, diabetes, thyroid disorder, infection, and antiphospholipid syndrome.

(B) Endometrial RIF
Presence of thin endometrium, less than 6 mm with impaired vascularity.

(C) Idiopathic RIF
Absence of factors from groups A and B. There is impaired immunological interaction between endometrium and embryos. It is in this group of idiopathic RIF, that immunotherapy can play an important role to help achieve a healthy pregnancy.

References

1. Christiansen OB, Nybo Andersen AM, Bosch E, et al. Evidence-based investigations and treatments of recurrent pregnancy loss. Fertil Steril. 2005;83(4):821–39.
2. The Practice Committee of the American Society for Reproductive Medicine. Evaluation and treatment of recurrent pregnancy loss: a committee opinion. Fertil Steril. 2012;98(5):1103–11.
3. American Society for Reproductive Medicines. Definitions of infertility and recurrent pregnancy loss. Fertil Steril. 2008;90(Suppl 5):S60.
4. Daya S. Habitual abortion. In: Copeland LJ, Jarrell JF, editors. Textbook of gynecology. 2nd ed. Philadelphia: W.B. Saunders; 2002. p. 227–71.
5. Mills JL, Simpson JL, Driscoll SG, Jovanic-Peterson L, Van Allen M, Aarons JH, et al. Incidence of spontaneous abortion among normal women and insulin-dependent women whose pregnancies were identified within 21 days of conception. N Engl J Med. 1988;319: 1617–23.
6. Carp HJ, Toder V, Torchinsky A, Portuguese S, Lipitz S, Gazit E, et al. Allogenic leukocyte immunization after five or more miscarriages. Recurrent Miscarriage Immunotherapy Trialistis Group. Hum Reprod. 1997;12:250–5.
7. The Practice Committee of the American Society for Reproductive Medicine. Evaluation and treatment of recurrent pregnancy loss: a committee opinion. Fertil Steril. 2012;98:1103–11.
8. Chard T. Frequency of implantation and early pregnancy loss in natural cycles. Baillieres Clin Obstet Gynaecol. 1991;5(1):179–89.
9. Roberts CJ. Letter: where have all the conceptions gone? Lancet. 1975;1(7907):636.
10. Kolte AM, van Oppenraaij RH, Quenby S, Christiansen OB, et al. Biochemical pregnancy loss and pregnancy of unknown location are prognostically important for unexplained recurrent miscarriage. Hum Reprod. 2014;29(5):931–7.

11. Royal College of Obstetricians and Gynaecologists. The investigation and treatment of couples with recurrent first-trimester and second-trimester miscarriage. RCOG Green-top Guideline No. 17. 2011.
12. van den Boogaard E, Cohn DM, Korevaard JC, et al. Number and sequence of preceding miscarriages and material age for the prediction of antiphospholipid syndrome in women with recurrent miscarriage. Fertil Steril. 2013;99:188–92.
13. Saravelos SH, Li TC. Unexplained recurrent miscarriage: how we explain it? Hum Rep. 2012;27(7):1882–6.
14. Rai R, Regan L. Recurrent miscarriage. Lancet. 2006;368(9535):601–11.
15. Nybo Andersen AM, Wohlfart J, Christens P, et al. Maternal age and fetal loss: population-based register linkage study. BMJ. 2000;320(7251):1708–12.
16. Li TC, Iqbal T, Anstie B, et al. An analysis of the pattern of pregnancy loss in women with recurrent miscarriage. Fertil Steril. 2002;78(5):1100–6.
17. Clifford K, Rai R, Regan L. Future pregnancy outcome in unexplained recurrent first trimester miscarriage. Hum Rep. 1997;12(2):387–9.
18. Stirrat GM. Recurrent miscarriage. Lancet. 1990;336:673–5.
19. Lathi RB, Gray Hazard FK, Heerema-Mckenney A, et al. First trimester miscarriage evaluation. Semin Reprod Med. 2011;29:463–9.
20. www.mayoclinic.orgart-20044134 (May 2016)
21. Clifford K, Rai R, Regan L. Future pregnancy outcome in unexplained recurrent first trimester miscarriage. Hum Reprod. 1997;12:387–9.
22. Christiansen OB, Pederson B, Rosgaard A, Husth M. A randomized, double-blind, placebo-controlled trial of intravenous immunoglobulin in the prevention of recurrent miscarriage: evidence for a therapeutic effect in women with secondary recurrent miscarriage. Hum Reprod. 2002;17:809–16.
23. Cowchock FS, Smith JB. Predicting for live birth after unexplained spontaneous abortions: correlations between immunological test results, obstetric histories, and outcome of the next pregnancy without treatment. AM J Obstet Gynecol. 1992;167:1208–12.
24. Nielsen HS, Christiansen OB. Prognostic impact of anticardiolipin antibodies in women with recurrent miscarriages negative for the lupus anticoagulant. Hum Reprod. 2005;20:1720–8.
25. Cauchi MN, Coulam CB, Cowchock S, Ho HN, Gatenby P, Johnson PM, et al. Predictive factors in recurrent spontaneous abortion-A multicentre study. Am J Reprod Immunol. 1995;33: 165–70.
26. Coughlan C, Ledger W, Wang Q, Liu F, Demirol A, Gurgan T, et al. Recurrent implantation failure: definition and management. Reprod Biomed Online. 2014;28(1):14–38.
27. Racowsky C. High rates of embryonic loss, yet high incidence of multiple births in human ART: is this paradoxical? Theriogenology. 2002;57(1):87–96.
28. Gnoth C, Godehardt D, Godehardt E, Frank-Herrmann P, Freundl G. Time to pregnancy: results of the German prospective study and impact on the management of infertility. Hum Reprod. 2003;18(9):1959–66.
29. Rinehart J. Recurrent implantation failure: definition. J Assist Reprod Genet. 2007;24(7):284–7.
30. Coughlan C, et al. Recurrent implantation failure: definition and management. BioMed Online. 2014;28(1):14–38.
31. Coulam CB. Implantation failure and immunotherapy. Hum Reprod. 1995;10(6):1338–40.
32. Tan BK, Vandekerckhove P, Kennedy R, Keay SD. Investigation & current management of recurrent IVF treatment failure in the UK. BJOG. 2005;112(6):773–80.
33. Thornhill AR, deDie-Smulders CE, Geraedts JP, Harper JC, Harton GL, Lavery SA, et al. ESHRE PGD Consortium 'Best practice guidelines for clinical preimplantation genetic diagnosis (PGD) & preimplantation genetic screening (PGS)'. Hum Reprod. 2005;20(1):35–48.

34. Margalioth EJ, Ben-Chetrit A, Gal M, Eldar-Geva T. Investigation and treatment of repeated implantation failure following IVF-ET. Hum Reprod. 2006;21(12):3036–43.
35. Alpha Scientists in Reproductive Medicine and ESHRE Special Interest Group of Embryology. The Istanbul consensus workshop on embryo assessment: proceedings of an expert meeting. Hum Reprod. 2011;26(6):1270–83.
36. Toth B, Wurfel W, Germeyer A, Hirv K, Makrigiannakis A, Strowitzki T. Disorders of implantation—are there diagnostic and therapeutic options? J Reprod Immunol. 2011;90(1): 117–23.
37. Timeva T, Shterev A, Kyurkchiev S. Recurrent implantation failure: the role of the endometrium. J Reprod Infertil. 2014;15(4):173–83.

Immune System

3

Abstract

The body's defense against foreign invaders is called the immune system. It protects us via a complex system that includes cells and chemicals (cytokines and chemokines) as well as organs within the immune system. The chapter discusses the normal immune response, types of immunity (innate and adaptive), the cells of the immune system, and the chemicals produced by them to induce the immune response. A brief explanation of the role played by Human Leucocyte Antigen (HLA) is also provided.

Keywords

Immune system · Immune response · Innate immunity · Adaptive immunity · Cells of immune system · Types of lymphocytes · T helper cells · T cytotoxic cells · T reg cells · NK cells · B lymphocytes · NKT cells · Neutrophils · Eosinophils · Basophils · Mast cells · Dendritic cells · Major histocompatibility complex · MHC · Cytokines

3.1 Introduction

The immune system is our body's protection against any invader. It defends the body against a wide range of pathogens including bacteria, viruses, and other microorganisms to alter cells of our own body. It is composed of cells, tissues, and organs that work together in defense against foreign invaders. The immune system has two jobs: to keep invaders outside of the body and to destroy them if they are already inside.

The function of the immune system has two activities, which are interrelated. These activities are:

M. Raut, M. Raut, *Lymphocyte Immunization Therapy (LIT) in Reproductive Failures*, https://doi.org/10.1007/978-981-19-2960-1_3

1. Recognition
2. Response

3.1.1 Recognition

The immune system can differentiate between self and non-self, that is, between the body's own antigens and the foreign invader antigens. It can differentiate between foreign molecules and the body's own molecules and cells. It can even detect the host cells that are altered and that can cause cancer. This ability of the immune system is called self/non-self-discrimination and is an important prerequisite for the protection it gives the body against foreign invasion.

3.1.2 Response

The recognition by the immune system leads to a response that eliminates or neutralizes the invader. This response is of two types:

1. Effector response
2. Memory response

3.1.2.1 Effector Response (See Fig. 3.1) [1]

The recognition of a foreign invader triggers the effector response by the immune system. It leads to elimination or neutralization of the invader. This is achieved through different types of cells and chemicals produced by these cells.

3.1.2.2 Memory Response (See Fig. 3.2) [2]

The other type of response to the foreign antigen is the memory response. The immune system can keep the foreign antigen in memory even after destroying it. This leads to a more rapid and heightened response when the invader attacks again. This property of memory protects us from the same disease a second time, and also helps us to recognize our own cells.

3.2 Types of Immunity

The immune system is composed of two different types (see Fig. 3.3) [3]:

1. Innate immunity
2. Adaptive immunity

These two systems collaborate together to give protection to our bodies.

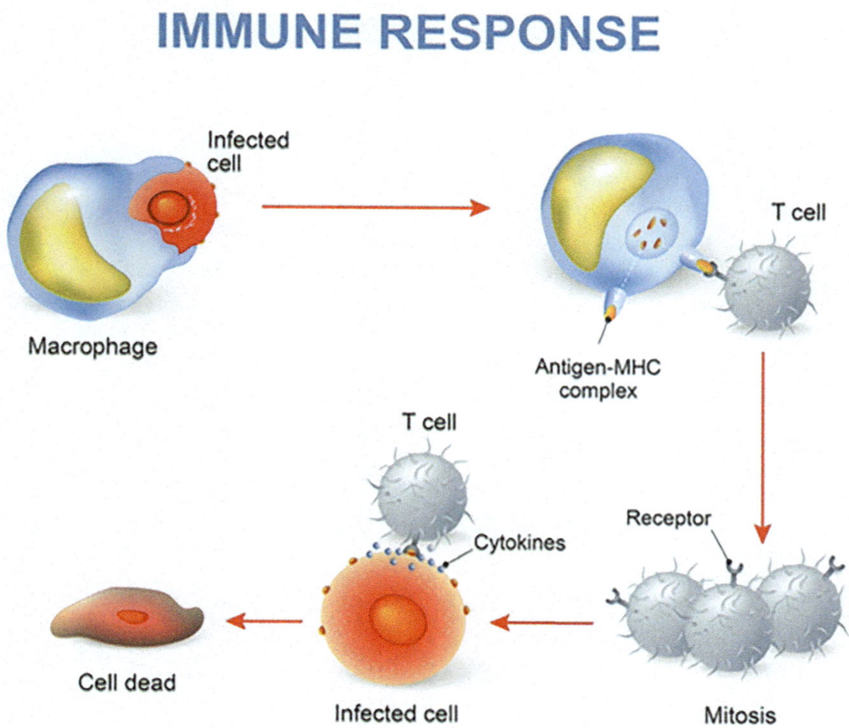

Fig. 3.1 Immune response to a foreign antigen (infection) [1]

3.2.1 Innate Immunity

Innate immunity includes the molecular and cellular mechanisms that are inherently present in the body, even before exposure to any foreign invader or infection. Innate immunity is not specific to any particular pathogen. It involves cellular and molecular components which acknowledge certain classes of molecules that are common to pathogens. Innate immunity can precisely distinguish between self and non-self, but it is unable to distinguish small differences in foreign molecules.

Innate immunity functions through cells and soluble molecules produced by cells. The cellular components are phagocytes. These cells are derived from monocytes, neutrophils, and tissue macrophages, and they act through a process of phagocytosis, which is a type of endocytosis involving the ingestion of extracellular particulate material and destroying it.

The soluble molecules involved in innate immunity are the protein lysosomes, the interferon proteins, and the complement system.

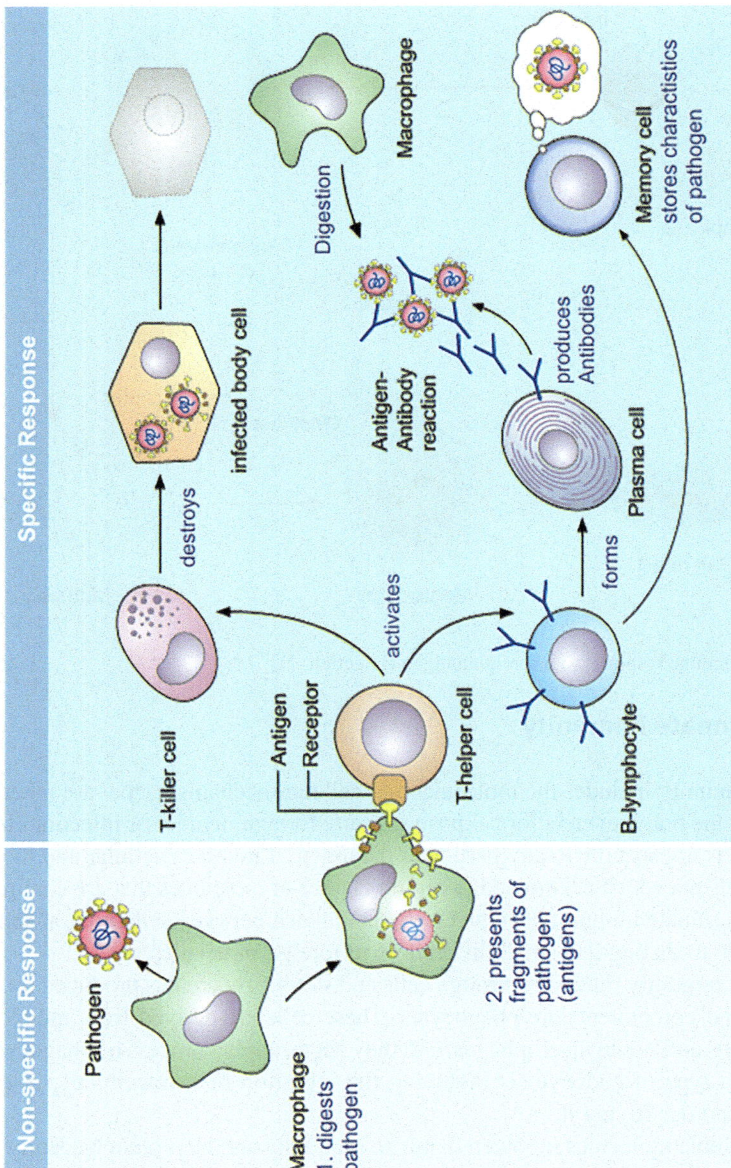

Fig. 3.2 Nonspecific and specific immune response (Cell Interplay During Nonspecific and Specific Immune Response) [2]

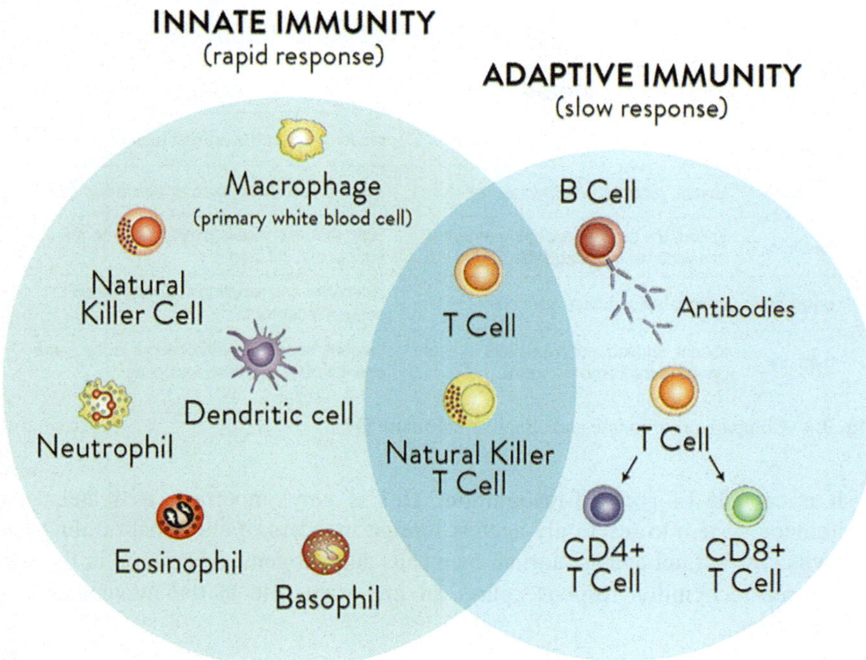

Fig. 3.3 Cells of innate and adaptive immunity [3]

3.2.2 Adaptive Immunity

This is the second type of immunity that evolves in reaction to infection or foreign antigen, and it alters itself to identify, eliminate, and then have a memory of the conquering pathogen or the foreign antigen.

Adaptive immunity starts developing a few days after exposure to the pathogen or the foreign antigen, and it provides the second line of defense. It also leads to the memory of the foreign antigen. This helps the adaptive immune system to make a rapid and more effective attack on the invader. Adaptive immunity functions through lymphocytes and the antibodies they produce.

The important features of adaptive immunity are as follows (see Fig. 3.4) [4]:

1. Antigen specificity:
   It can distinguish subtle differences among antigens.
2. Diversity:
   Due to this, it can recognize a single type of organism and differentiate among those with minor generative variations.
3. Immunologic memory:
   Due to this, the second exposure to the same antigen or pathogen leads to heightened response. This can lead to lifelong immunity against many infections.
4. Non-self-recognition:

Innate *vs* adaptive immunity

	innate	adaptive
self / non-self discrimination	present, reaction is against foreign	present, reaction is against foreign
lag phase	absent, reponse is immediate	present, response takes at least a few days
specificity	limited, the same response is mounted to a wide variety of agents	high, the response is directed only to the agents that initiated it.
diversity	limited, hence limited specificity	extensive, and resulting in a wide range of antigen receptors.
memory	absent, subsequent exposures to agent generate the same response	present, subsequent exposures to the same agent induce amplified reponses

Fig. 3.4 Comparison of innate and adaptive immunity [4]

It is capable of non-self-recognition. This is very important, as it helps the immune system to react only against foreign invaders of abnormal or abnormal own cells, but not against normal own cells and antigens. Aberration in the self/non-self-recognition response plays an important role in the maintenance of pregnancy.

3.3 The Cells of the Immune System

The immune system functions through a wide range of cells. Lymphocytes bearing antigen receptors are the most important cells of the immune system. There are other types of white blood cells that also play essential roles in generating and affecting the immune response (See Fig. 3.5) [5].

3.4 Lymphocytes

Lymphocytes make up 20–40% of white blood cells in the body. Lymphocytes move in blood and lymph. They also migrate into tissues and lymphoid organs.

There are three major subtypes (see Fig. 3.6) [6]:

1. B lymphocytes
2. T lymphocytes
3. Natural killer cells

Different types of lymphocytes are recognized by different membrane molecules. These membrane molecules can be recognized by monoclonal antibodies, which are antibodies that recognize a specific epitope of an antigen. The collection of a monoclonal antibody that reacts with a specific membrane molecule is organized together as a Cluster of Differentiation (CD). These CD molecules are also known as

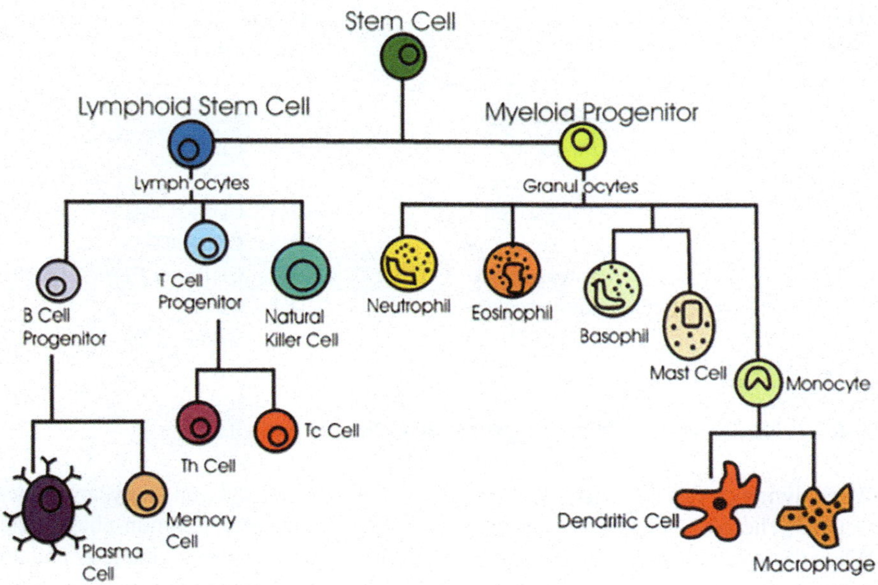

Fig. 3.5 Cells of the immune system [5]

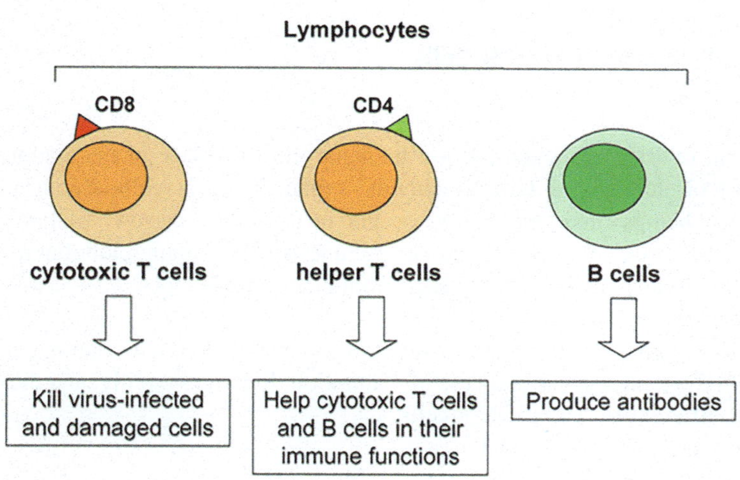

Fig. 3.6 Types of lymphocytes [6]

CD markers. There are about 250 CD markers known to exist (see Table 3.1 and Fig. 3.7).

Table 3.1 Common CD markers

Type of cell	CD markers
Stem cells	CD34+, CD31−
All leukocyte groups	CD45+
Granulocyte	CD45+, CD15+
Monocyte	CD45+, CD14+
T lymphocyte	CD45+, CD3+
T helper cell	CD45+, CD3+, CD4+
Cytotoxic T cell	CD45+, CD3+, CD8+
B lymphocyte	CD45+, CD19+ or CD45+, CD20+
Thrombocyte	CD45+, CD61+
Natural killer cell	CD16+, CD56+, CD3

3.4.1 Naïve and Effector Lymphocytes

When lymphocytes T or B have not interacted with an antigen, they are called "naive." These naive lymphocytes are small, 6 μm in diameter, and non-phagocytic. When these cells interact with antigens, they enlarge up to 15 μm in diameter and are called lymphoblasts. The lymphoblasts multiply and differentiate to become memory cells and effector cells. It is these effector cells that help to eliminate antigens in different ways depending on the type of lymphocytes.

3.4.2 B Lymphocytes (B Cells)

1. Origin:
   B cell, or B lymphocyte, has got its name from its place of maturation, in the bursa of fabricius, a lymphoepithelial organ found in gallinaceous birds. In human beings, their maturation occurs in the bone marrow. Mature B cells synthesize and display specific membrane-bound immunoglobulin molecules by which they can be distinguished from other lymphocytes.
2. Maturation:
   When a naive B cell binds to the antigen that matches its membrane-bound antibody, it undergoes a process of maturation. It divides rapidly, and the divided cells differentiate into effector cells.
3. Two types: (a) Plasma cells
   (b) Memory B cells
   (a) Plasma cells:
   Plasma cells produce antibodies that can be secreted. Plasma cells do not divide further and are highly specialized for the secretion of antibodies, capable of secreting up to 1000 molecules of antibody per second. They do not have membrane-bound antibodies.
   (b) Memory B cell:
   Memory B cells have a membrane-bound antibody. Persistence of these cells is responsible for lifelong immunity to many pathogens.

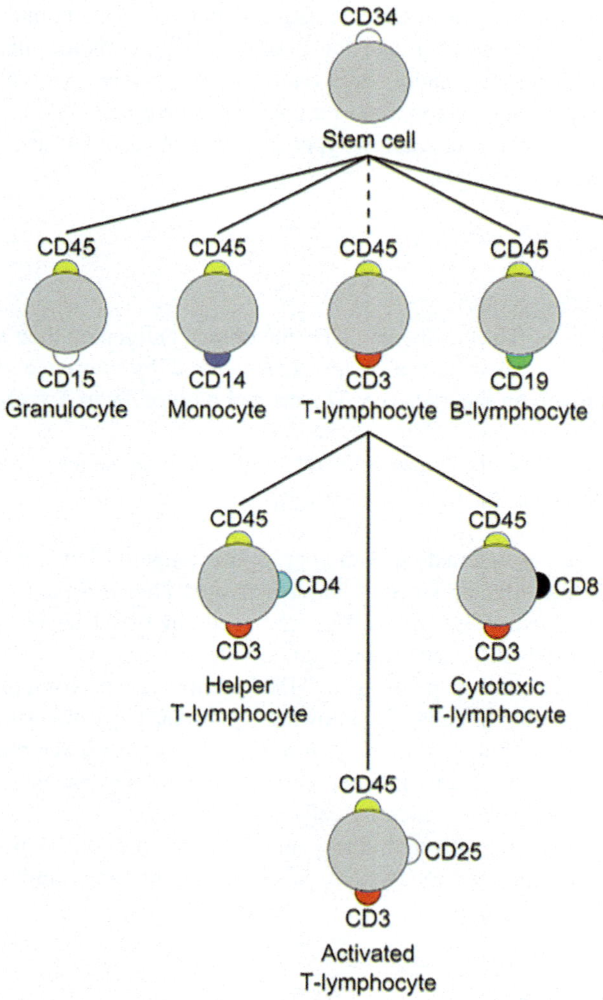

Fig. 3.7 CD markers on lymphocytes

3.4.3 T Lymphocytes (T Cells)

1. Origin:

   T lymphocytes are named after their site of formation, the Thymus. During the process of maturation in the thymus, the T cell expresses unique antigen-binding molecules on its membrane and is called T cell receptor.

T cell receptors only acknowledge antigens that have been bound to cell membrane proteins known as major histocompatibility complex molecules (MHC molecules). There are mainly two types of MHC molecules: Class one MHC molecules that are evident in almost all nucleated cells and class two MHC molecules that are evident in only a few types of cells that carry out the function of antigen presentation.

2. Types:
   (a) T helper (Th) cells
   (b) T cytotoxic [Tc] cells
   (c) T regulatory [Treg] cells
   (a) T helper (Th) cells: Th cells display CD4 membrane antigen on their surface.
   (b) T cytotoxic (Tc) cells: Tc cells are characterized by presence of CD8 membrane antigen on their surface. The normal ratio of Th to Tc cells is 2:1 in the body.
   (c) T regulatory (Treg) cells: These cells are identified by presence of CD4 and CD25 membrane antigens on their surface.
3. Maturation:
   Th cells get activated by interaction with appropriate antigen, MHC complexes, and get converted into effector Th cells. These activated Th cells help to activate B cells, DC cells, macrophages as well as other cells involved in the immune response. Some Th cells differentiate into memory cells.

   When a Tc cell recognizes the antigen, MHC complexes, it proliferates and differentiates into effector Tc cells, also known as cytotoxic T lymphocyte (CTL). The CTL eliminates anything that displays foreign antigen complexed with class one MHC (e.g., tumor cell, virus infected cell, or cells of foreign tissue graft). Some Tc cell lines can become memory cells.

   When T reg cells are activated, they suppress the immune response by negatively regulating the immune system. Some T reg cells turn into memory cells.

3.4.4 Natural Killer Cells

1. Introduction:
   Natural killer (NK) cells are larger granular lymphocytes and have cytotoxic activity against a wide range of tumor cells and virus infected cells. They constitute 5–10% of lymphocytes in the peripheral blood.

   NK cells make up a portion of the innate immune system. They lack the membrane molecules and receptors found in B and T cells. Human NK cells have been identified as $CD^{3-}CD56^{+}CD16^{+}$ or $CD^{3-}CD56^{+}CD16^{-}$. $CD56^{+}$ can be $CD56^{bright}$ or $CD56^{dim}$. The $CD56^{bright}$ are mainly in the peripheral blood. $CD56^{dim}$ are present in the endometrial lining. $CD56^{dim}$ cells can maturate into $CD57^{+}$ cells. The $CD57^{+}$ NK cells are highly cytotoxic and are detrimental to pregnancy. They can cause RPL or RIF [7].
2. Mechanism of action:

(a) Use of NK cell receptors: NK cells use NK cell receptors to identify irregularities such as decreasing visibility of class one MHC molecules or an abnormal surface antigen on tumor cells or viruses infected cells. The cells are then destroyed.
(b) Antibody-dependent cell mediator cytotoxicity (ADCC): Tumor cells and virus infected cells have antitumor or antiviral antibodies bound to their surface due to the immune response of the body. NK cells, due to the membrane receptor CD16 can attach to these antibodies and destroy these cells.

3.4.5 NKT Cell

NKT cells have features of T cells and NK cells. They have T cell receptors like T cells, but unlike the T cell, they interact with MHC-like molecules. Like NK cells, they can kill target cells. Some of the NKT cells can secrete the cytokines needed by B cells to produce antibodies and are required for the development and expansion of cytotoxic T cells.

3.5 Mononuclear Phagocytes

1. Introduction:
   The mononuclear phagocytic system is made up of monocytes and macrophages in the blood and tissues. Monocytes flow through the bloodstream for around 8 h, and it is during this time that they expand and then move into the tissues, differentiating into particular tissue macrophages. They remain in a particular tissue and are called fixed macrophages, or they may be free or wandering.
2. Activation:
   Macrophages are activated by different stimuli. Phagocytosis of a particular particulate antigen is in contact with receptors sensing molecules in pathogens that lead to activation. They are also activated by cytokines secreted by activated TH cells.
3. Action:
   Activated macrophages lead to phagocytosis of microbes, increase secretion of inflammatory mediators, and cultivation of TH cells. They release cytotoxic proteins that can kill the cells infected with a virus, tumor cells, and intracellular bacteria. The antigen is converted into peptides during the phagocytic process. These peptides are attached to class two MHC molecules forming complexes. These peptide-class two MHC complexes are displayed at macrophage membranes. This helps in Th cell activation that is required for humoral and cell-mediated immunity.

4. Types:
   (a) *Granulocytic cells:*
      (i) Neutrophils:
      Neutrophils, also known as granulocytes, are granulocytes with a mildly lobed nucleus. They have granulated cell cytoplasm that can be stained with acid and basic dyes. They are produced in the bone marrow. From there they are released into the peripheral blood. They circulate in the blood for 7–10 h, then move into the tissues. When there is infection in the body, bone marrow releases neutrophils into the circulation. Action: The chemotactic factors at the inflammatory site lead to accumulation of neutrophils. These chemotactic factors are cytokines secreted by activated Th cells, components of the blood clotting system, and some complement components. Neutrophils destroy the pathogen by phagocytosis.
      (ii) Eosinophils:
      Eosinophils cells have a bilobed nucleus, granulated cell cytoplasm that stains with acid dye used in red. They are important in the defense against parasitic organisms. They secrete eosinophilic granules that damage the parasitic membranes. They have less phagocytic activity.
      (iii) Basophils:
      Basophils are characterized by lobed nucleus and highly granulated cell cytoplasm. They can be stained with methylene blue, a basic dye. They produce active substances that can trigger allergic reactions from their granules. They do not have any phagocytic activity.
   (b) *Mast cells*:
   The bone marrow produces precursors of mast cells that are then released into the blood. They differentiate into mast cells. They are able to enter the tissues and have large quantities of cytoplasmic granules that contain histamine and various other active substances. They are important in the development of allergies.
   (c) *Dendritic cells*:
   The immune system's first cells to be discovered were the dendritic cells. Dendritic cells are characterized by long membranous extensions that resemble dendrites in nerve cells. There are four types:
      (i) Langerhans DC: Found in skin.
      (ii) Interstitial DC: Present in interstitial spaces of all organs except the brain.
      (iii) Monocyte-derived DC: Derived from monocytes migrated into tissues.
      (iv) Plasmacytoid-derived DC: Arrives from plasmacytoid cells. They play an important role in innate immunity. They also act as antigen-presenting cells.

   DC has importance in the immune system. They monitor the pathogens or foreign antigens and capture them. Then they present them to Th cells.

3.6 Major Histocompatibility Complex

3.6.1 Introduction

Major histocompatibility Complex (MHC) is a set of genes that play an important role both in intercellular identification and in distinguishing self from non-self. It participates in the development of humoral as well as cell-mediated immune responses. It is called the human leukocyte antigen, HLA, complex in humans.

3.6.2 Structure

The MHC is present with the long continuous stretch of DNA on chromosome six. The HLA complex encodes three types of glycoprotein molecules in humans:

1. Class one MHC gene: Glycoproteins that are present on the surface cells of all nucleated bodies belong to the class one MHC gene.
2. Class two MHC genes: Include glycoproteins present on antigen-presenting cells macrophages, dendritic cells and B cells, and has three regions, DP, DQ, and DR.
3. Class three MHC genes: Include proteins having immune function (components of complement system and inflammation process).

3.6.3 Function

1. Class I MHC genes—Presentation of peptide antigens to Tc cells (CD8+).
2. Class II MHC genes—Presentation of peptide antigens to Th cells (CD4+).
3. Class III MHC genes—Secretion of proteins of complement system and molecules involving inflammation.

HLA-DQ Alpha in class two MHC plays an important role in the process of allorecognition of pregnancy and subsequent acceptance or rejection of pregnancy.

When the HLA-DQ genes on the trophoblast and those on the maternal T cells are diverse, there is easy immune recognition that helps in the acceptance of pregnancy. However, with sharing of HLA-DQ genes, this allo-recognition is lost leading to immune response that causes rejection of pregnancy.

3.7 Cytokines

3.7.1 Introduction

Immune responses are mediated by a set of proteins secreted from immunologically active cells. These proteins are known as cytokines. Cytokines are responsible for the regulation of immune effector cells development, and some of them have direct effector functions.

Cytokines include those proteins which are secreted by lymphocytes (lymphokines) as well as by monocytes and macrophages (monokines). Many cytokines are referred to as interleukins, as they are secreted by leukocytes. Cytokines also include chemokines that affect chemotaxis.

3.7.2 Functions

1. Building of cellular as well as humoral immune responses.
2. Initiation and maintenance of inflammatory responses.
3. Control of cellular multiplication and differentiation.
4. Wound healing.

3.7.3 Types

1. According to type of immunity
   (a) Cytokines of innate immunity: IL-1, IL-6, IL-12, TNF-α, IFN-α, IFN-β
   (b) Cytokines of adaptive immunity: IL-2, IL-4, IL-5, Transforming growth factor TGF-β, IFN-γ
2. According to type of response
   (a) Cytokines with Th1 response: IL-2, IFN-γ, TNF-β, GM-CSF, IL-3
   (b) Cytokines with Th2 response: IL-3, IL-4, IL-5, IL-10, IL-13

3.7.4 Importance

Cytokines produced by Th1 cell substrates promote its growth and inebriates the development and activity of Th2 cell substrates. The same is true for cytokines produced by Th2 cell substrates. This is called cross regulation. This mechanism plays an important role in immunological maintenance or rejection of pregnancy. Loss of balance between Th1 and Th2 responses can lead to alloimmune miscarriage with Th1 responses overcoming Th2 responses.

References

1. https://reverehealth.com/live-better/brief-overview-immune-system/
2. http://www.intactschools.eu/node/361
3. https://quizlet.com/361084423/blood-Figcells-and-the-immune-system-diagram/
4. https://www.pinterest.com/pin/141652350755381380/
5. Todar's Online Textbook of Bacteriology.
6. https://demo.oppia-mobile.org/media/courses/cd3-et/01_15073_en.html
7. Nielsen CM, White MJ, Goodler MR, Riley EM. Functional significance of CD57 expression on human NK cells and relevance to disease. Front Immunol. 2013;4:422.

Immunology of Pregnancy (Reproductive Immunology)

4

Abstract

This chapter on reproductive immunology explains the normal immune response during pregnancy. The fetus is antigenically different from the mother (50%) and is a semi-allograft. In spite of this, it is able to survive successfully in the mother's body ("Paradox of pregnancy"). The various immune mechanisms that may protect the pregnancy have been discussed in the chapter in detail. The current evidence appears to favor the role of Treg cells in protecting the pregnancy. Th1/Th2 balance and NK cells are also important in maintenance of pregnancy. This will help to understand the underlying alloimmune mechanisms, responsible for "unexplained RPL and RIF."

Keywords

Allograft · Paradox of pregnancy · Pregnancy immune response · Theories of immune tolerance · Bipolar immune response · Network hypothesis · Th1/Th2 response · Treg/Th17 response · Allorecognition · NK cells · Pre-implantation immune suppression · eu-FEDS hypothesis · Mechanism of immune protection/rejection

4.1 Introduction

Recurrent miscarriages can be a devastating event in a woman's life. Infertility is also a problem, especially after repeated failed IVF attempts. These recurrent miscarriages can be caused by a number of factors. Recurrent miscarriages are common in 0.5–1% of the population.

Out of these couples, 50% are those who have done all of the investigations necessary to rule out the possibility of a known cause. These cases are then labelled as unexplained miscarriages or unexplained implantation failures [1]. It has been shown that a significant number of unexplained RM/RIFs have been linked to

M. Raut, M. Raut, *Lymphocyte Immunization Therapy (LIT) in Reproductive Failures*, https://doi.org/10.1007/978-981-19-2960-1_4

alloimmune conditions, over the past 25 years. It has also been seen that more the number of miscarriages, the more is the chance of alloimmune factors.

To understand the alloimmune rejection of pregnancy, it is essential to know about the immunological response in normal pregnancy.

4.2 Fetus as Allograft

Pregnancy is Nature's Miracle. This is apparent when you look at the immunological aspects of pregnancy.

A pregnancy has 50% genetic material from the father and 50% from the mother. The paternal part of the fetus carries with it the paternal antigens. These antigens are totally foreign to the mother's immune system. So, logically speaking, the mother's immune system should build an immunological response against the fetus, leading to the rejection of fetus [2]. Thus, the fetus is a semi-allograft in the immunologically active maternal immunological system.

The fetus survives and grows into a healthy baby, despite this. This is known as the "Paradox in pregnancy," and it has always been a problem for reproductive immunologists.

4.3 Immune Response in Normal Pregnancy

1. Bipolar Immune Response
   As the fetus is a semi-allogenic graft, an immunological response is immediately triggered. This immune response can be either positive or negative. These two responses coexist at the same time and are known as the Bipolar Immune Reaction.

   The harmful response is the rejection reaction against the paternal antigens in the fetus. This response is both humoral and cell mediated. The humoral response is by cytotoxic antibodies and the cell-mediated response is through cytotoxic cells. The favorable response is the facilitation reaction. This response is predominantly humoral. It counteracts the harmful response of rejection. In the case of pregnancy, there is an immune rejection response by default. Only when this is counteracted by the facilitation response, there is any sustenance to pregnancy [19, 20].
2. Network Hypothesis (Niels Jerne, 1974)
   The Network Hypothesis which Niels Jerne proposed in 1974 about the mechanisms of regulation of our immune response can explain why there is a favorable immunological response.

   Any antibody has a unique set of antigenic determinants (epitopes) and it is called an idiotope. It is the actual antigen binding site. If a separate antibody is produced that has specific binding capabilities similar to an idiotope of the antibody, then it is called an anti-idiotope antibody. This anti-idiotypic antibody then binds with the B Lymphocyte receptor for the original antigen instead of the original antibody and inhibits the immune response to the antigen.

Table 4.1 Types of cytokines

Type	Cytokines
Th1 (pro-inflammatory)	IL-1beta, IL-2, Interferon gamma (IFN-gamma), IL-6, IL-7, IL-12, IL-15, IL-17, IL-18,IL-21,TNF-alpha
Th2 (anti-inflammatory)	IL-4, IL-5, IL-10, IL-13, IL-35, TGF-beta, GM-CSF

3. Th1 Response Versus Th2 response (see Table 4.1)
   Most biological effects in the immune system are caused by Cytokines. T Lymphocytes can be subdivided into two groups based on the type of surface molecules they contain, known as CD4 or CD8. They are also the major source of cytokines. The T lymphocytes that contain CD4 are called helper T cells, and they can be divided further into Th1 and Th2 and the cytokines they make are called Th1- and Th2-type cytokines, respectively (see Table 4.1).

   Th1 cytokines produce pro-inflammatory responses. Examples of Th1 cytokines are IFN-γ and IL2. The excessive pro-inflammatory response causes uncontrolled tissue damage. The hypothesis is that Th1 cytokines can cause miscarriage through a process called "cytokine-triggered vascular autoamputation." This activates the coagulation system and can cause vasculitis, thrombosis, and affects the blood supply to the embryos. Th2-type cytokines produce anti-inflammatory responses. A few examples of them are IL4, IL5, IL3, and IL10. Th2 responses counteract the Th1-mediated destructive responses [3, 4].

   The role of cytokines in the growth of pregnancy was first put forward by Wegmann in 1987 in his "Immunothrophic Theory" [5, 6]. In 1993, he further suggested that Th1/Th2 equilibrium is important in pregnancy. There is a change in the Th1/Th2 proportion during pregnancy. This means that Th2-type cells are dominant over Th1-type cells.

   This is good for the embryo as it increases the growth and function of the placenta and prevents reaction to the trophoblasts. Certain molecules in trophoblast (HLA-G) and certain proteins produced by trophoblast (Heat Shock Proteins (HSP), pregnancy-specific beta1 glycoprotein) help to change toward predominant Th2 response in pregnancy.

   HCG produced by trophoblast also plays a role in enhancing Th2 response. HCG induces the production of progesterone by the corpus luteum. Progesterone causes the production of a protein called progesterone-induced blocking factor (PIBF). This immunoregulatory protein leads to the production and release of IL-4 from T Lymphocytes, thereby increasing the Th2 response.
4. Concept of Regulatory T cells (Treg) and Inflammatory Th Cells (Th17)
   There are two more important immune factors at the feto–maternal interface which help in the Th1/Th2 equilibrium. These are—Regulatory T cells (Treg) and Inflammatory T cells (Th17).

   Treg cells can be described as a subset of immune-regulatory T lymphocytes. They are described as CD4+CD25+ FOXP3+ lymphocytes. They are produced by activation of naive CD+ T cells in the thymus. These cells promote immune tolerance by inhibiting cytokine production and proliferation by T cells, and antibody production by B cells. They also inhibit the maturation process of dendritic cells and suppress the cytotoxic activity of natural killer (NK) cells. This is achieved through "Bystander Suppression Phenomenon." Treg cells secrete inhibitory cytokines (IL-10, TGF-β, and IL-35) and suppress the release of pro-inflammatory cytokines (IL-1b and IL-6, TNFa, IL-22). Treg cells also have a cytolytic effect on pro-inflammatory T cells and dendritic cells through the release of granzymes and perforin [7–10].

TGF-1β, IL-1β, and IL23 can be used to differentiate Th17 cells from naive T-cells. They produce effecter cytokines IL17, IL-17F, IL-21, and IFN-γ. Presence of Th17 cells leads to loss of maternal tolerance to pregnancy [11–13].

The interaction between Th17 cells and Treg cells shows the delicate balance between the facilitator and suppressor responses and the eventual outcome of pregnancies.

5. Concept of Allorecognition and Immune Tolerance—Role of NK Cells
The innate immune system includes natural killer (NK) cells. They are important in controlling the immune response at feto–maternal interaction.
   (a) Peripheral NK cells [14, 15]
   The levels of peripheral blood NK cells in normal pregnancy are within normal limits. However, there are many studies that suggest that the levels of peripheral blood NK cells are higher in women aborting karyotypically normal pregnancies. The CD56+ NK cells in peripheral blood with increased expression of CD94 and decreased expression of CD69 are associated with healthy pregnancy.

   About 10–15% of peripheral blood lymphocytes are made up of NK cells with CD16 and CD56 as their cell surface antigens. Based on CD56 expression intensity, NK cell populations are split into two groups.
      (i) CD56dim with high expression of CD16
      (ii) CD56bright with low levels of CD16 or no CD16

   CD56dim is more cytotoxic. CD56bright is the main source of dNK cell-derived immune-regulatory cytokines. During normal pregnancy, peripheral NK cells decrease in number.

   Peripheral NK cells are also regulated by hormones. Estrogen acts by direct action on Estrogen receptors. Progesterone acts by action on progesterone receptors through the production of PIBF. Thus, there is hormonal regulation of peripheral NK cell activity protecting pregnancy.
   (b) Decidual NK cells (dNK cells) [14]
   From the onset of pregnancy, the dominant cell population is the decidual NK (dNK) cells. Mostly, dNK cells are CD56 bright16-. These cells are vital for normal pregnancy.

   The blastocyst implantation and decidualization activate dNK cells. Other events at the Feto–Maternal Interface depend on the allorecognition, which occurs at decidua.
   (c) NK cell Allorecognition System.

After implantation, the invading trophoblastic cells express certain HLA-class 1 molecules (mainly HLA G or C). The dNK cells have receptors that immunologically interact with these molecules on the trophoblast. This interaction between trophoblastic and NK cells has resulted in the creation of the NK Cell Allorecognition System.

The interaction of NK cell receptors with antigens of the trophoblast can provide two types of signals:

(i) Activation of cytotoxic NK cells (Th1 response)
(ii) Inhibition of dNK activation (Th2responce)

Of these two responses, which response will predominate depends on the allorecognition process. If there is allorecognition, i.e., the inhibitory dNK receptors recognize the specific ligands on the trophoblast, then there will be inhibition of dNK cell activation for trophoblast damage. If this allorecognition does not happen, then dNK cells cause destruction of the trophoblast. In this allorecognition process, inhibitory KIR receptors (inhKIR) on dNK cells and HLA G and C play an important role [16].

6. Concept of Pre-implantation immune Suppression [17, 18]
Other than Treg cells and Th2 cytokines, there are many other mechanisms that can induce immune tolerance in semi-allogenic fetuses through the maternal immune system. These are known as preimplantation immunosuppressive factors.
   (a) Sperm: Sperms may produce local immunosuppression by the release of prostaglandins.
   (b) Seminal plasma: TGF-β present in the seminal plasma may send signal for producing Granulocyte Macrophage Colony Stimulating Factor (GM-CSF) and the accumulation of Treg cells in the uterus.
   (c) Effect of circulating hormones: Progesterone can induce immunosuppression at decidua through PIBF, even before implantation takes place.
   Estrogen can act through Estrogen Receptors to decrease NK cell activity.
   (d) Cytokines released during ovulation and fertilization produce immunosuppressive effect on deciduas.
7. Eutherian Fetoembryonic Defense System (eu-FEDS) hypothesis
The eu-FEDS hypothesis assumes that both soluble and cell surface-associated glycoproteins that are expressed on gametes, suppress any possible immune responses and inhibit rejection of a fetus. It also proposes that there are certain carbohydrate sequences called oligosaccharides that are covalently linked to immunosuppressive glycoproteins and serve as functional groups, suppressing the immune response. Alpha-fetoproteins, CA125, glycodelin-A, and Placental Protein 14 (also known as PP14) are the key fetal and maternal glycoproteins implicated in the eu-FEDS human model.

4.4 Mechanisms of Immunoprotection of Embryo

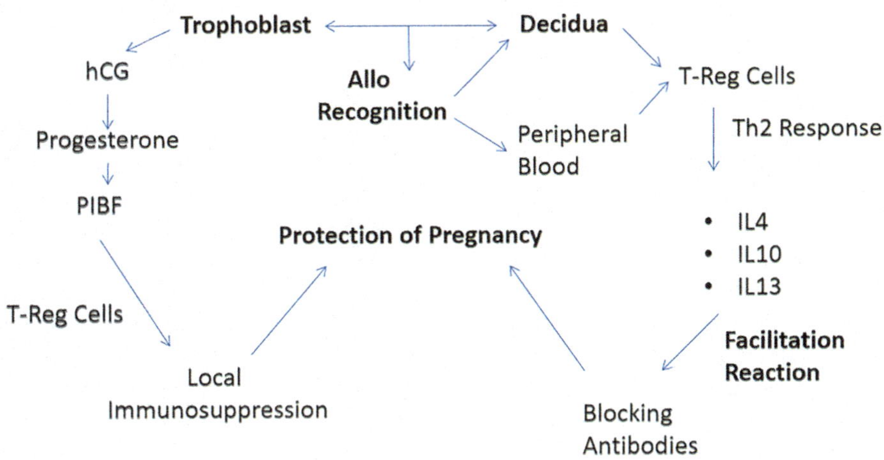

4.5 Mechanisms of Alloimmune Rejection of Pregnancy

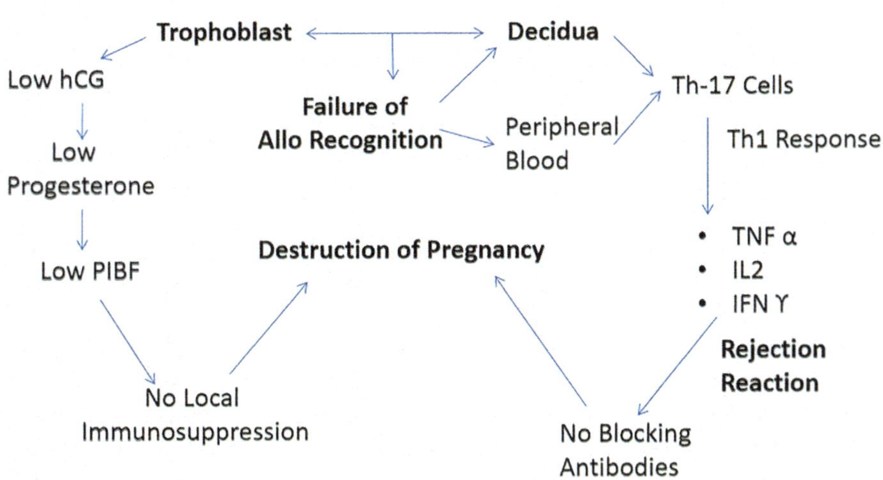

References

1. McIntyre JA, Coulam CB, Faulk WP. Recurrent spontaneous abortion. Am J Reprod Immunol. 1989;21:100–4.
2. Labarrere CA. Allogenic recognition and rejection reaction in placenta. Am J Reprod Immunol. 1989;21:94–9.
3. Chaouat G, Tranchot Diallo J, Volumenie JL, et al. Immune suppression and Th1/Th2 balance in pregnancy revisited: a (very) personal tribute to Tom Wegmann. Am J Reprod Immunol. 1997;37:427–34.
4. Raghupathy R. Th1-type immunity is incompatible with successful pregnancy. Immunol Today. 1997;18:478–82.
5. Wegmann TG. Placental immunotrophism: maternal T cells enhance placental growth and function. Am J Reprod Immunol. 1987;15:67–9.
6. Wegmann TG, Lin H, Guilbert L, et al. Bidirectional cytokine interactionsin the maternal-fetal relationship: is successful pregnancy a Th2 phenomenon? Immunol Today. 1993;14:353–6.
7. Szekeres-Bartho J, Barakonyi A, Polgar B, et al. The role of gamma/delta T cells in progesterone-mediated immunomodulation during pregnancy: a review. Am J Reprod Immunol. 1999;42:44–8.
8. Saito S, Nakashima A, Shima T, et al. Th1/Th2/Th17 and regulatory T-cell paradigm in pregnancy. Am J Reprod Immunol. 2010;63:601–10.
9. Bluestone JA, Abbas AK. Natural versus adaptive regulatory T cells. Nat Rev Immunol. 2003;3:253–7.
10. Sakaguchi S. Naturally arising CD4+ regulatory T cells for immunologic self-tolerance and negative control of immune responses. Annu Rev Immunol. 2004;22:531–62.
11. Jonuleit H, Schmitt E, Kakirman H, et al. Infectious tolerance: human CD25+ regulatory T cells convey suppressor activity to conventional CD4+ T helper cells. J Exp Med. 2002;196:266–0.
12. Aluvihare VR, Kallikourdis M, Betz AG. Regulatoty T cells mediate maternal tolerance to the fetus. Nat Immunol. 2004;5:271–88.
13. Arruvito L, Sanz M, Banham AH, et al. Expansion of CD4+CD25+ and FOXP3+ regulatory T cells during the follicular phase of menstrual cycle: implications for human reproduction. J Immunol. 2007;178:2572–8.
14. Nielsen CM, White MJ, Goodler MR, Riley EM. Functional significance of CD57 expression on human NK cells and relevance to disease. Front Immunol. 2013;4:422).
15. Seshadri S, Sunkara SK. Natural killer cells in female infertility and recurrent miscarriage. Hum Reprod. 2014;20(3):429–38.
16. Varla-Leftherioti M. Role of a KIR/HLA-C allorecognition system in pregnancy. J Reprod Immunol. 2004;62:19–27.
17. Robertson SA, Guerin LR, Bromfield JJ, et al. Seminal fluid derives expansion of the CD4+CD25+ T regulatory cell pool and induces tolerance to paternal alloantigens to mice. Biol Reprod. 2009;80:1036–45.
18. Robertson SA. Seminal plasma and male factor signaling in the female reproductive tract. Cell Tissue Res. 2005;322:43–52.
19. Voisin GA. Immunological facilitation, a broadening of the concept of the enhancement phenomenon. Prog Allergy. 1971;15:328–485.
20. Chaouat G, Voisin GA, Escalier D, et al. Facilitation reaction (enhancing antibodies and suppressor cells) and rejection reaction (sensitized cells) from the mother to the paternal antigens of the conceptus. Clin Exp Immunol. 1979;35:13–24.

Diagnosis of Alloimmune Factor

5

Abstract

The success of the treatment of RPL/RIF depends greatly on the diagnosis. Even after doing all the investigations to find out the cause of RPL/RIF, in almost 50% of cases, the underlying cause cannot be pinpointed. These cases are labeled as "unexplained" RPL/RIF. It is in these patients, that the underlying cause can be alloimmune rejection of pregnancy. In this "unexplained" group, the presence of alloimmune factors can be found by certain immunological tests, which have diagnostic value. However, more standardization of these tests is required.

Keywords

Diagnosis of RPL · Diagnosis of RIF · Genetic factor · Karyotype · PGT · Autoimmunity · Thrombophilia · Lymphocyte crossmatch · NK cells · TNF-alpha · CD57+ cells · Treg cells · HLA DQ · KIR genotype · Th1:Th2 ratio · CD4: CD8 ratio

5.1 Introduction

2–5% of couples in their reproductive years experience repeated miscarriages [1, 2].

Similarly, 10% of couples undergoing IVF suffer from Recurrent IVF failure [3]. These conditions can be very detrimental to the patient's health, relationships with their spouse, and quality of life. Around 50% of patients diagnosed with RM do not have an identifiable cause. They are called unexplained RM. RIF is also a common cause of RM that remains undiagnosed.

Research has been extensive in order to discover the mechanism behind these "unexplained" RMs and RIFs. These cases are consistent with an immunological mechanism being the primary cause [4]. These groups of patients with URM and URIF appear to be mainly associated with alloimmune factors. It is in these groups that immunomodulatory treatments play an important role. While most of the

M. Raut, M. Raut, *Lymphocyte Immunization Therapy (LIT) in Reproductive Failures*, https://doi.org/10.1007/978-981-19-2960-1_5

research on immunologic causes has been done, there is still controversy over how to treat these patients.

RPL treatment is only possible if the cause of the problem is identified. This can only be achieved after a thorough evaluation of the patient to determine the underlying cause of RPL. It is extremely difficult to treat the patient effectively if a complete investigational workup of the patient is not done.

The diagnosis of alloimmune factors can be done based on Clinical evaluation and Laboratory investigations.

1. Clinical evaluation:
   (a) Clinical history:
   A complete workup of a patient with RPL or IF includes a complete history. This is to include a review of past obstetric history. This should incorporate the following:
      (i) Number of miscarriages.
      (ii) Dates.
      (iii) Duration of pregnancy.
      (iv) Spontaneous or induced.
      (v) Fetal cardiac activity seen or not.
      (vi) Evidence of subchorionic bleeding.
      (vii) Presence of missed abortion.
      (viii) Whether D & C is done.
      (ix) Investigations done during pregnancy.
      (x) Karyotype of POC.
      (xi) The history should also include any previous disease, acute or chronic infection, any previous gynecological or other surgeries.
      (xii) Family history of pregnancy loss.
      (xiii) History of blood transfusion.
      (xiv) Smoking and alcohol.
      (xv) History of consanguinity.

   It cannot be stressed enough how important it is to have a detailed and accurate history to determine the underlying cause of RPL.
   (b) Clinical examination:
   An integral part of any evaluation is a detailed clinical examination. So, it is important to analyze all previous investigations.
2. Laboratory investigations:
   Successful immunotherapy relies on the selection of the right patient with RPL/IF. This is possible only after doing all the tests that will rule out all possible causes other than the alloimmune factor. This will be followed by immunological testing for alloimmune problems.

5.2 Level One

This includes all the investigations that are done to rule out the known causes of RPL/RIF. A brief list of these tests is given below. The details of these investigations are out of the scope of this book.

5.2.1 Test for Genetic Factors

Genetic causes of Recurrent Miscarriage are an important cause of sporadic miscarriage but not so common in RPL.

1. Recurrent miscarriage can be caused by changes in the maternal or paternal genes. These changes can be:
   (a) Karyotype abnormalities
   (b) Recessive and dominant carrier status
   (c) Gene mutations responsible for coagulations and metabolism of folates
2. Genetic Abnormalities in the Fetus.
   Two ways in which a fetus might have abnormal chromosomes:
   (a) Recurrent aneuploidy means the total number of chromosomes is abnormal (too many or too few).
   (b) Abnormalities in one of the parent's chromosomes are repeatedly passed onto the baby in each successive pregnancy. These are usually structural abnormalities or rearrangements in one of the parent's chromosomes (translocations).

   If the fetus was chromosomally healthy, then it is possible that miscarriage could be due to an alloimmune factor.

   Kwinecka-Dmitriew et al. found a lower rate of chromosomal abnormalities in material from repeated miscarriages than isolated miscarriages, leading to the conclusion that other factors could cause recurrent miscarriages [5].

Diagnosis
1. Tests of POC:
   Classic karyotype
   CGH
   MLPA
   QF-PCR
   BoBs
   FISH with probes
   NGS
2. Karyotype of parents:
   Karyotype
   Molecular studies (DNA): RPL profile—to detect genes that can predispose to RPL.
3. Preimplantation genetic diagnosis:

(a) PGT-A (PGS): Preimplantation genetic diagnosis for aneuploidy. This is done to detect abnormal chromosome numbers.
(b) PGT-M (PGD): Preimplantation genetic diagnosis. This is used to identify genetic defects in embryos during IVF to prevent miscarriage or implantation failure.

5.2.2 Test for Anatomical Factors

1. USG 2D or 3D.
   Sonography helps to detect anatomical abnormalities in the uterus that can lead to miscarriages.
   (a) Uterine fibroids, especially submucous fibroids or fibroid polyps.
   (b) Congenital anomalies of the uterus, e.g., septate and sub-septate uterus, bicornuate uterus, and arcuate uterus.
   (c) Intrauterine adhesions (Asherman syndrome).
   Sonography helps to detect the blood flow to the endometrium and uterine blood flow.
   (a) Endometrial blood flow: Reduced flow predisposes to RPL.
   (b) Uterine blood flow: Increased uterine blood flow resistance can lead to RPL.
2. HSG.
3. Laparoscopy and hysteroscopy.
4. Test for incompetent cervix.

5.2.3 Test for Endocrine Factors

A. FBS, PLBS, HbA1c.
B. Thyroid disorders, T3, T4, and TSH. Antithyroglobulin antibodies, antiTPO antibodies.
C. Test for PCOD: USG, FSH/LH, AMH, Serum testosterone, Serum DHEA, Serum insulin.
D. Adrenal disorders.
E. Luteal phase defect, serum progesterone, and EB.

5.2.4 Test for Infections

(a) Torch profile, (b) Chlamydia antibodies, (c) Tuberculosis.

5.2.5 Test for Autoimmunity

(a) ANA, (b) dsDNA, (c) Anti-cardiolipin antibodies, (d) Anti-complement antibodies, (e) PTTK-Lupus anticoagulant, (f) Antibody against phosphatidylserine, phosphatidylethanolamine, phosphatidylcholine, phosphatidylglycerol, phosphatidic acid, phosphatidylinositol, (g) Antibody against cofactor proteins, prothrombin, annexin U, anti-ovarian antibody, and antiβ2GPI antibodies.

5.2.6 Test for Thrombophilia

1. *Congenital Thrombophilia*
   Factor V Leiden
   Prothrombin G20210A substitution mutation.
   Hereditary protein C deficiency and protein S deficiency.
   Antithrombin deficiency. Anti-thrombin III activity.
   Homozygous MTHFR mutation—C677T and A1298C.
   Hereditary factor F8 deficiency.
   Fibrinogen deficiency.
   Plasminogen Activator Inhibitor PAI-1(4G/5G) polymorphism.
2. *Acquired Thrombophilia*
   (a) Antiphospholipid antibodies against phosphatidylserine, phosphatidylethanolamine, phosphatidylcholine, phosphatidylglycerol, phosphatidic acid, and phosphatidylinositol.
   (b) Anticardiolipin antibodies.
   (c) PTTK and Lupus anticoagulant.
   (d) Anti-Beta2 glycoprotein I.

5.2.7 Other Tests

(1) Serum homocysteine, (2) Histone antibodies, (3) Sjogren's antibody. SSA, SSB, (4) Scl-70 antibodies, (5) Serum prolactin.

5.3 Level Two

Tests for alloimmune factor:

1. Lymphocyte crossmatch or LAD, estimation of APCA or LAD, leukocyte antibody detection panel.
2. Peripheral NK cell.
   (a) Complete assay.
   (b) NK cell inhibition panel.

3. uNK cells.
4. Serum TNF-alpha.
5. HLADQ-alpha genotype.
6. T-regulatory cell Treg estimation.
7. Th1:Th2 intracellular cytokine ratio.
8. KIR genotyping.
9. CD4:CD8 ratio.

5.3.1 Lymphocyte Crossmatch

Synonyms: Leukocyte antibody detection—LAD, HLA crossmatch, Anti-paternal leukocyte antibody detection.

Introduction

LAD is a comprehensive test that helps to determine the reason for reproductive failure. It allows you to identify if there is an alloimmune cause of RM. It measures how much antibody reacts with a protein that is not self (paternal or donated) lymphocytes.

Principle

Alloimmune recognition plays a very important role in this by the formation of Blocking Antibodies. These antibodies were measured in a variety of assays. They are called Anti-paternal Cytotoxic Antibodies(APCA), Asymmetric Antibodies, or Mixed Lymphocyte Reaction Blocking Antibodies (MLRBA) [6–8]. The level of asymmetric antibodies is marginally lower in women who have suffered pregnancy loss [9].

Numerous studies have shown a link between repeated abortions and parental similarity in Human Leukocyte Antigens. This causes hyporesponsiveness and inhibits the production of blocking antibodies that act as immunological regulators to preserve the pregnancy. Anti-idiotopic Antibodies, Mixed Lymphocyte Reaction Blocking Antibodies, and Anti-paternal Cytotoxic Antibodies are all part of the regulators that shield the paternal HLA molecules present on the exterior of the fetus. They also act as a barrier to attack the maternal T cells or NK cells [10].

Lymphocyte crossmatch or HLA crossmatch helps in the detection of anti-paternal antibodies in the mothers' blood.

Methods for Lymphocyte Crossmatch

- Complement-Dependent Cytotoxicity (CDC) method—By Serology
- Flow Cross Match (FCXM)—By flow cytometry.
- Virtual Cross Match/ Luminex XM—By Donor-Specific Antibody (DSA) assay using Luminex platform.

Results

CDC Crossmatch

This is the commonest method used to detect the presence of anti-paternal antibodies. This is done manually using the Complement Dependent Cytotoxicity method. It requires viable lymphocytes. In this method, the wife's serum is incubated with the husband's/donor's lymphocytes. The results are interpreted as follows:

Results for lymphocyte crossmatch

	Result
Reaction of patient's serum (Neat) with donor's lymphocytes (HLA crossmatch)	Negative/Positive
Reaction of patient's serum (1:2) with donor's lymphocytes	Negative/Positive
Reaction of patient's serum (1:4) with donor's lymphocytes	Negative/Positive
Reaction of patient's serum with patient's lymphocytes (Autoantibodies)	Autoantibodies detected/ not detected

Interpretation of the results

Interpretation	% Dead lymphocytes	[a]Scoring system
Negative	0–10	1+
Doubtful negative	11–20	2+
Weakly positive	21–40	4+
Positive	41–80	6+
Strongly positive	81–100	8+

[a]Scoring system: Percentage of cell death and interpretation are recorded as per ASHI guidelines

Positive LAD patients can be further tested with luminex anti-HLA testing to identify which HLA antibodies are reacting with the husband's WBC. The test is considered false positive if there are no anti-paternal HLA antibodies. This could be due to infection or other immune complexes.

Importance

A negative lymphocyte crossmatch test in presence of unexplained RPL or RIF indicates an alloimmune cause and that immunomodulation (LIT) will be beneficial in such a case. The lymphocyte crossmatch may not be always positive in women who have spontaneous successful pregnancies without a history of RPL or RIF.

5.3.2 Peripheral Blood NK Cells (pNK Cells)

Introduction

The establishment of early pregnancy is possible thanks to the crucial role played by NK cells. They can be found in blood, all tissues, and even the uterine endometrium. RM and RIF can be caused by NK cell dysfunction.

Principle
Estimation of pNK cells can help to establish the possibility of alloimmune cause of RM or RIF. Also, pNK cell levels correspond to the levels of uNK cells.

Mechanism
pNK cells are a part of the innate immunity and are responsible for immune surveillance. They act on the cells, which are not allorecognized, killing them. They do not require activation.

Types
NK cells are an integral part of the natural immune system. They do not have the membrane molecules or the receptors found in T cells and other B cells. Human NK cells can be identified as $CD3^{-}CD56^{+}CD16^{+}$ or $CD3^{-}CD56^{+}CD16^{-}$. $CD56^{+}$ can be $CD56^{bright}$ or $CD56^{dim.}$ The $CD56^{bright}$ are mainly in the peripheral blood. $CD56^{dim}$ are present in the endometrial lining. $CD56^{dim}$ cells can mature into $CD57^{+}$ cells. The $CD57^{+}$ NK cells are extremely cytotoxic and harmful to pregnancy. They can be linked to RPL, RIF, or unexplained infertility [11].

Method
Flow cytometry.

Results
Different assays.

A. Proportion of pNK cells out of all the lymphocytes.
   Proportion of pNK Cells: normal value—less than 18%.
B. Concentration.
   Concentration of activated $CD56+^{dim}$ cells. Determined with CD69 marker.
      Normal value: Less than $12 \times 10^{6}/l$.
C. In vitro assay NK cell inhibition.
   Inhibition of NK cell activity.

Interpretation
A. Proportion of NK cells
   Normal range varies with each lab.
   CD3: 60–85%
   CD19: 2.0–12.0%
   CD56: 2.0–12.0%
   CD19+ cells: 8.0–10.0%
B. NK cell inhibition panel.
   IVIG and intralipid are used to test inhibition of NK cells at various dilutions.
      Normal range varies with each lab.
   Normal percentage:
   50: 1 dilution - 10.0–40.0%

25: 1dilution - 5.0–30%
12.5: 1 dilution - 3.0–20%
15–25% of women suffering from RM have high NK cell levels [12].

C. K-562 target cell test (for NK cell activity)
This test can be used to determine the cytotoxic activities of human natural killer (NK) cells. To distinguish between target and effector cells, the K562 target cells are labeled with a lipophilic fluorescent membrane dye. After incubation, dead target cells can be identified using a DNA stain. This penetrates dead cells and stains their nuclei. This allows the determination of the percentage of target cells killed by effector NK cells. Flow cytometry is used to determine NK activity [13].

Importance
Increased concentration or proportion of NK cells and abnormal NK cell inhibition assay indicates the presence of the alloimmune factor. These cases benefit from immunotherapy.

A meta-analysis of 22 studies done by Seshadri and Sankara showed that:

- Increased pNK cells number in UI and RM patients
- Increased pNK cells percentage in RM patients [14]

Future Prospects
Better methods, interpretation, and larger studies are required.

5.3.3 Uterine NK Cells

Introduction
In the implantation process of early pregnancy, the NK cells in the uterine endometrium play a critical role. It has been demonstrated that excessive activity and concentration of uNK cells can be linked to RM and RIF.

Origin of Cells
uNK cells are a part of innate immunity. Majority of uNK cells are recruited from the peripheral blood NK cells. Their concentration is 10% in the proliferative phase, 20% in late secretory phase, and more than 30% in early pregnancy.

Human NK cells do not express the membrane molecules and the receptors seen in T cells and B cells. They have been identified as $CD^{3-}CD56^{+}CD16^{+}$ or $CD^{3-}CD56^{+}CD16^{-}$. $CD56^{+}$ can be $CD56^{bright}$ or $CD56^{dim}$. The $CD56^{bright}$ are mainly in the peripheral blood. $CD56^{dim}$ are present in the endometrial lining. $CD56^{dim}$ cells in the endometrial lining can maturate into $CD57^{+}$ cells. The CD57+ NK cells are highly cytotoxic and are detrimental to pregnancy. They can cause RPL or RIF [11].

Mechanisms

uNK cells interact with trophoblast cells and may regulate the physiological invasion of trophoblast by allorecognition. Overactivity of uNK cells can cause RM or RIF by direct cytotoxicity of cytokine production.

Principle

Estimation of uNK cells can help to establish the possibility of alloimmune cause for RM or RIF.

Methods

uNK cells are assessed by taking an endometrial biopsy and testing it for the presence of uNK cells and their concentrations.

Timing

EB is done between 19th and 23rd day of the cycle.

Collection of Sample

The endometrial biopsy material is collected in formalin (60%), that is diluted by 50% (1:1 with water).

Method of Testing

A. Immunohistochemistry
   Easier method
   Can only count CD56 and CD57 cells
   Measurement of subtypes or level of activation is not possible.
   A reliable and consistent method of counting is required.
B. Flow Cytometry
   It is technically difficult but can assess the surface makers, CD56+ bright/ CD56+ dim.

Results

Positive: CD57 cells present.
Negative: Absence of CD57 cells.
15–25% of RM high uNK cells.

Interpretation

This test has limitations because endometrial tissue is complex with cell clusters and uNK cell counts may vary to a great extent, and the uNK cell numbers change rapidly daily in the luteal phase.

Importance

Increased uNK cell CD56 or presence of CD57 indicates a higher association with alloimmune cause for RIM or RIF.

Future Prospects

uNK cell estimation is a highly specialized investigation. A dedicated and interested pathologist is required for the accuracy of results. Daily reference ranges for a given population are required to be prepared. This will help in better interpretation of results.

5.3.4 Serum TNF-alpha

Introduction

Tumor Necrosis Factor (TNF-alpha) is a cytokine involved in systemic inflammation. TNF-alpha plays an important role in the maintenance of pregnancy and in alloimmune rejection leading to RM or RIF.

Principle

TNF-alpha is a Th1 type of cytokine. The different cells in the endometrium (lymphocytes including NK cells, fibroblasts, and other immune cells) express TNF-alpha and is required for a variety of normal endometrial functions. It is pro-inflammatory and elevated levels are associated with RM or RIF.

Mechanism

TNF-alpha is produced by the activated macrophages and NK cells. Elevated levels have deleterious effects on oocyte, implantation of embryo, and future growth of embryo by inducing inflammatory reaction. Along with other pro-inflammatory cytokines like IFN-γ, it causes apoptosis of corpus luteal cells causing luteal insufficiency and subsequent pregnancy loss.

Method

Plain blood collected and serum is separated. Done by chemiluminescent assay.

Results and Interpretation

Normal level: 4–12 ng/ml.

Raised levels of TNF-alpha with a history of uRPL/uRIF indicate alloimmune problem and immunomodulatory treatment (LIT) will be beneficial. In the studies conducted by Daher S. et al. and Kim H. R. and colleagues, notably higher levels of TNF-alpha were found in RPL patients than in controls [15, 16].

Importance

Elevated levels of serum TNF-alpha are associated with RM or RIF or IVF failure. Even after treatment with immunotherapy, the success rates are limited in the presence of elevated serum TNF-alpha. Levels of TNF-alpha should be brought to normal with medications before planning pregnancy or IVF cycle.

Future Prospects

Anti-TNF-alpha medications can help normalize TNF-alpha levels. These medications can also cause immunosuppression which makes the patient more susceptible to other diseases. Better anti-TNF-alpha medicines in future will definitely help to overcome this problem, which can be treated effectively.

5.3.5 HLADQ-alpha Genotype

Human leukocyte antigen (HLA) system is an important system, responsible for our unique immune system. It helps our body to recognize what is our own and what is foreign immunologically.

HLADQ-alpha genotype is a DNA test that determines the HLA typing. White blood cells have class II HLA on their surface. These can be of HLADR or HLA-DQ. Each tissue type is composed of two parts: alpha A and beta B. HLA-DQ Alpha [DQA1] and/or DQ Beta [DQB1] mothers are more likely to miscarry before the eighth week of their pregnancy, according to research.

Mechanism

If a couple shares HLADQ-Alpha loci, then they will have the same protein sequences at the cell surface. The resulting embryo also has identical protein sequences on their cell surface. The mother's immune system believes that the embryo is a part of her, but it should not be there, just like in the case of cancer cells. Hence, the natural killer cells [NK cells] in the uterine lining overproduce inflammatory cytokines (Th1 type) and attack the trophoblastic cells of the embryo, leading to implantation failure or miscarriage. It has also been shown that compatibility of HLADQ-alpha between mother and fetus can lead to exacerbation of autoimmune disorders like RA and APLA syndrome, leading to increased incidence of miscarriage.

Method

The estimation of HLA is done by polymerase chain reaction (PCR) using blood.

5.3.6 T-reg Cell Estimation (CD4+CD25+foxP3)

T-regulatory cells (T-reg) are a subset of T lymphocytes that have potent suppressive activities and play an important role in preventing autoimmune diseases and cutting down immune responses.

Apart from Th1/Th2 responses, Th17 and T-reg cell responses are also important. T17 cells produce pro-inflammatory cytokine IL-17, leading to inflammatory response. In contrast, T-reg cells play an important role in immunoregulation and induction of tolerance. T-reg cells prevent proliferation and cytokine release in CD4+ andCD8+ T cells. These cells produce immunoglobulin and prevent cytotoxic activity from natural killer NK cells. They also mature dendritic cells (DC), which results in tolerance induction [11, 12].

Good levels of T-reg cells are essential to prevent production of autoantibodies (preventing autoimmune disease). They are important for reproductive immune tolerance to allow the embryos to implant (preventing IVF failure), and reduce the risk of miscarriage.

Testing
T-reg (CD4 CD25 fox P3+)
Normal limit: More than or equal to 0.7%.

5.3.7 TH1-TH2 Intracellular Cytoplasmic Ratio

Introduction
This is a test to determine the balance between the Th1 [pro-inflammatory] and Th-2 [anti-inflammatory] immune response and its possible role in repeated miscarriages or implantation failure.

Principle
Th1 cells are part of cell-mediated immunity and release various cytokines in the response to foreign proteins. It produces a pro-inflammatory response. Th2 cells mediate antibody-mediated response and it is anti-inflammatory. The test determines, which response is dominant.

Mechanism
Th-1 cells produce cytokines interferon-γ, Il-1B, TNF-α, leading to inflammatory response. Th2 cells produce cytokines IL4, IL-5, IL-6, IL-10, and IL-13 leading to anti-inflammatory response.

Method
Luminex processing for Th1 cytokines—[IL-2, IFN-gamma, TNF-alpha], Th2 cytokines—[IL4, IL5, IL10]. Fasting is not required. T helper cells get stimulated and produce cytokines. This indicates whether they are Th1- or Th2-type cells. The cells are then treated with antibodies to detect the cytokines. Flow cytometry allows for the identification of individual cells

Results and Interpretation
The three states are (see Table 5.1):

A. Balanced
B. Strong Th1 dominance
C. Strong Th2 dominance

Cytokine panel level 1: IL1, IL6, IL10, TNF-alpha.

Table 5.1 Cytokines and Th dominance

Cytokines	TH1 dominance	TH2 dominance
IL2	High	Low
IL4	Low	High
IL12	High	Low
IFN-gamma	High	Low
TNF-alpha	High	Low

Cytokine panel level 2: IL1, IL2, IL4, IL5, IL6, IL10, IL13, IFN-gamma, TNF-alpha, TNF- beta, TGF-beta.

Th1: Th2 cytokine ratios can also be studied.

Normal limits
TNF-alpha:IL10 - 13.2–30.3
IFN-gamma:IL10 - 5.6–21.0

Raised ratios indicate an inflammatory response.

Importance
Th1 dominance indicates an alloimmune factor responsible for repeated miscarriages or implantation failure.

Future Prospects
Tests are useful if an experimental treatment is done and in future can be useful for clinical diagnostic applications.

5.3.8 KIR Genotyping

Natural killer (NK) cells are lymphocytes belonging to the natural immune response system. They are important in maintaining normal pregnancy by preventing immunological rejection of embryos.

The interaction between surface Killer cell Immunoglobulin-like Receptors (KIR) and class I HLA determines the functional activity of NK cells. NK cells can have several types of KIR on their surfaces. These receptors are either inhibitory or activating in nature.

Fourteen KIR genes are located on chromosome 19q13.4, with two pseudogenes. These 16 KIR gene combinations are expressed by human NK cell haplotypes. Group A has fewer inhibitory receptors and group B has a greater number of activating receptors.

KIRs may activate NK cells or inhibit their production of soluble factors required for pregnancy maintenance. These KIRs are believed to be responsible for recurrent miscarriages [17].

Varla Leftherioti et al. proposed that the cause of the miscarriage in women who had experienced an alloimmune abortion is that they have a limited number of inhibiting KIR repertoire. Hence, the decidual NK cells do not recognize the trophoblastic molecule belonging to class 1 and perceive them as foreign [18].

In RPL couples, if there is a parallel between male HLA-C2 or female HLA-C1 ligand KIR receivers, it could disturb the balance of activatory and inhibitory KIR ligand interactions when pregnancy occurs. The HLA–KIR interactions may cause disruption in individual immune recognition. There are many possible mechanisms for that [19].

Method

5 ml blood is collected in EDTA, EDTA-coated collection vials taken. DNA extraction is done. DNA sample typed for KIR genes responsible for inhibitory signals and activating signals. Multiplex PCR-SSP reactions are used.

5.3.9 CD4/CD8 Ratio

Introduction

In this test, the percentage of CD4 cells, T helper cells, and CD8 cells, deregulatory cells, are counted and compared. It can give an indication about the presence of an alloimmune problem in RM-RIF.

Principle

CD4-positive T lymphocytes are helper cells and CD8-positive T lymphocytes are suppressor cells. An increase in CD8 percentage with the resultant decreased CD4 to CD8 ratio indicates alloimmune dysfunction.

Results and Interpretation

Normal CD4:CD8 ratio is 0.54–2.97.

Low CD4:CD8 ratio is associated with RPL/RIF.

The proportion of CD8+ was much more in women with RM than it was in the control ($p < 0.05$). The CD4:CD8 ratio in women with RM was lower. The CD4/CD8 ratio was lower in women with RM compared with the control group. These observations support the concept that an increase of CD8(+) lymphocytes could be involved in the etiology of RPL [20].

References

1. Daya S. Habitual abortion. In: Copeland LJ, Jarrell JF, editors. Textbook of gynecology. 2nd ed. Philadelphia: W.B. Saunders; 2002. p. 227–71.
2. Mills JL, Simpson JL, Driscoll SG, Jovanic-Peterson L, Van Allen M, Aarons JH, et al. Incidence of spontaneous abortion among normal women and insulin-dependent women whose pregnancies were identified within 21 days of conception. N Engl J Med. 1988;319: 1617–23.
3. Coughlan C, et al. Recurrent implantation failure: definition and management. BioMed. Online. 2014;28(1):14–38.

4. Takeshita T. Diagnosis and treatment of recurrent miscarriage associated with immunologic disorders: is paternal lymphocyte immunization a relic of the past? J Nippon Med Sch. 2004;71: 308–13.
5. Kwinecka-Dmitriew B, Zakrzewska M, Latos-Bieleńska A, et al. Częstość występowania aberracji chromosomowych w materiale z poronień. Ginekol Pol. 2010;81:896–901.
6. Pandey MK, Thakur S, Agrawal S. Lymphocyte immunotherapy and its probable mechanism in the maintenance of pregnancy in women with recurrent spontaneous abortion. Arch Gynecol Obstet. 2004;269:161–72.
7. Ito K, Tanaka T, Tsutsumi N, et al. Possible mechanisms of immunotherapy for maintaining pregnancy in recurrent spontaneous aborters: analysis of anti-idiopathic antibodies directed against autologous T-cell receptors. Hum Reprod. 1999;14:650–5.
8. Kishore R, Agarwal S, Halder A, et al. HLA sharing, anti-paternal cytotoxic antibodies and MLR blocking factors in women with recurrent spontaneous abortion. J Obstet Gynaecol Res. 1996;22:177–83.
9. Barrientos G, Fuchs D, Schrocksnadel K, et al. Low levels of serum asymmetric antibodies as a marker of threatened pregnancy. J Reprod Immunol. 2009;79:201–10.
10. Higuchi K, Aoki K, Kimbara T, Hosoi N, Yamamoto T, et al. Suppression of natural killer cell activity by monocytes following immunotherapy for recurrent spontaneous aborters. Am J Reprod Immunol. 1995;33:221–7.
11. Nielsen CM, White MJ, Goodler MR, Riley EM. Functional significance of CD57 expression on human NK cells and relevance to disease. Front Immunol. 2013;4:422).
12. Sacks G. NK cells in peripheral blood and the endometrium. In: Christiansen OB, editor. Recurrent pregnancy loss. West Sussex: Wiley; 2014.
13. Kandarian F, Sunga GM, Arango-Saenz D, Rossetti M. A flow cytometry-based cytotoxicity assay for the assessment of human NK cell. Activity J Vis Exp. 2017;126:56191.
14. Seshadri S, Sunkara SK. Natural killer cells in female infertility and recurrent miscarriage. Hum Reprod Update. 2014;20(3):429–38.
15. Daher S, Camano L. Cytokines in recurrent pregnancy loss. J Reprod Immunol. 2004;62(1-2): 151–7.
16. Kim HR, Park AJ, Lee MK. CD56/CD16 expression on mononuclear cells and concentration of serum TNF-alpha in recurrent spontaneous abortion. Korean J Lab Med. 2006;26(3):198–203.
17. Ozturk OG, Sahın G, Karacor EDZ, Kucukgoz U. Evaluation of KIR genes in recurrent miscarriage. J Assist Reprod Genet. 2012 Sep;29(9):933–8.
18. Varla Leftherioti M, Spyropoulou-Vlachumm M, Keramitsoglout T, et al. Lack of appropriate natural killer cell inhibitory receptors in women with spontaneous abortion. Human immunology. 2005;66:65–71.
19. Elbaşı MO, et al. Maternal killer-cell immunoglobulin-like receptors and paternal human leukocyte antigen ligands in recurrent pregnancy loss cases in Turkey. Clin Exp Reprod Med. 2020;47(2):122–9.
20. Ghafourian M, Abuhamidy A, Karami N. Increase of peripheral blood TCD8(+) cells in women with recurrent miscarriage. J Obstetr Gynaecol. 2014;34:36–9. https://doi.org/10.3109/01443615.2013.817980.

Part II

Lymphocyte Immunization Therapy

Introduction to Lymphocyte Immunization Therapy (LIT)

6

Abstract

Lymphocyte immunization therapy (LIT) is an active immunomodulatory treatment used in "unexplained" reproductive failures, where the cause is likely to be alloimmune rejection of pregnancy. Many studies support the efficacy of LIT while others are not in favor of LIT. This part of the book examines all the aspects of LIT to assess its efficacy and its potential role in future in the management of reproductive failures.

Keywords

LIT · Lymphocytes · Efficacy of LIT · Future of LIT

Lymphocyte Immunization Therapy (LIT) is the first immunomodulatory treatment used to treat unexplained RM patients. As the name suggests, it involves the immunization of the prospective mother with allogenic mononuclear cells, mainly lymphocytes. LIT is performed with paternal and donor lymphocytes. Lymphocytes are the most important cells in the immune system. They play an important role both in cell-mediated immune responses and in humoral immune responses. They are the key players in rejection response against immunologically different substrates. Hence, the use of lymphocytes has been suggested to modulate the immune response against immunologically different fetuses in the mother's body (semi-allograft).

LIT has been in practice since 1981 when it was introduced for the first time. Different medical professionals have developed their own methods and protocols to achieve different success rates. While many studies support the efficacy and validity of LIT, others have questioned its efficacy. The mode of action and safety have been extensively discussed in different studies. LIT has also gone through some controversies.

We will therefore examine all aspects of LIT and evaluate its parameters to get a clear picture of the current state of LIT. This will allow us to assess the effectiveness

M. Raut, M. Raut, *Lymphocyte Immunization Therapy (LIT) in Reproductive Failures*, https://doi.org/10.1007/978-981-19-2960-1_6

of LIT and how to choose the couples that will benefit from LIT. This information will allow us to evaluate the potential role LIT could play in the future of uRM/uRIF.

History of LIT

7

Abstract

Lymphocyte Immunization Therapy (LIT) was the first immunomodulatory treatment suggested for repeated pregnancy loss (RPL). It was Sir Peter Medawar, who first proposed that the maternal immune system has a big part to play in maintaining pregnancies that are immunologically different. The first LIT was used by Taylor and Faulk. However, it was Dr. Beer, who put forward the concept that failure of maternal recognition of paternal alloantigens could cause deficient tolerance to the pregnancy and that paternal mononuclear cellular immunization may improve maternal recognition of paternal alloantigen, which could prevent recurrent loss of pregnancy (RPL). Further studies by Mowbray, RMITG, Daya, and Carp established the efficacy of LIT.

Keywords

History of LIT · Reproductive immunology · HLA sharing · Blocking antibodies

7.1 History

Pregnancy has a semi-allograft nature and it is, therefore antigenically different from the mother. The mechanisms involved in implantation prevent the allograft from being rejected. If the immunological tolerance becomes impaired, then spontaneous abortion may occur.

The immunological basis of pregnancy was always considered by researchers and various theories were put forward to explain the mechanisms that help in maintenance of the semi-allograft, i.e., pregnancy. Once the alloimmune basis of pregnancy was established, various immunological treatments were put forward to treat complications of pregnancy, resulting from the disturbances in alloimmune imbalance. A possibility of immunization of paternally marked cells was explored to counter the immune response of the mother toward her paternally registered fetus.

M. Raut, M. Raut, *Lymphocyte Immunization Therapy (LIT) in Reproductive Failures*, https://doi.org/10.1007/978-981-19-2960-1_7

One of the first modalities of immunomodulatory treatment that was tried and recommended was Lymphocyte Immunization Therapy (LIT).

When we look at the history and contributions of LIT, the following researchers and their studies are major contributors:

Dr. Peter Medawar:

7.2 Sir Peter Medawar (1915–1987)

Sir Peter Medawar, who is considered the father of Reproductive Immunology, was the first to suggest the role of immunology in the maintenance of pregnancy. He first put forward the question, "How does a pregnant mother manage to nourish within herself, for many weeks and months, a fetus that is an antigenically foreign body?" The immune system is our body's protection against any invader. It defends the body against a wide range of pathogens including bacteria, viruses, other microorganisms, and altered cells of our own body. To defend against foreign invaders, there is a network that includes cells, tissues, organs, and other parts. Yet, this powerful immune system is overcome in mammals in pregnancy where the fetus grows in the mother's body.

Dr. Medawar states that there are immunological and endocrinological issues in vertebrates that were caused by the evolution in viviparity, and certain immunological changes can play an important part in maintaining pregnancy. He suggested that whether it is due to physical separation by the placenta or by other means, the maternal immune response must be insensitive to the fetus' presence in some way. We now know that this theory is not the correct one, but it made the reproductive immunologists think about the "Paradox of pregnancy" [1].

7.3 Dr. Billingham R.E.

Billingham was the first person to describe lymphocyte immunization. Billingham et al. conducted experiments on how to make tissue homografts immunologically acceptable to hosts, which would normally react against them. This phenomenon is called "actively acquired tolerance." This acquired tolerance is immunologically specific and is due to a specific failure of the host's immunological response.

Billingham RE et al. observed that skin transplants among fraternal twins were accepted, but not between non-twin siblings. They suggested that blood cells exchanged via the twin placentas continued and maintained an immunologic tolerance [2].

7.4 Dr. Taylor C and W. Page Faulk (1981)

This study was based on the hypothesis that it should be possible to artificially stimulate protective factor production in the mother by infusing her with allogenic leukocytes, thereby, protecting the fertilized egg at implantation.

Four women who had experienced three spontaneous abortions tested positive for A, B, C, and DR histocompatibility (HLA). They also shared the same antigens as their husbands. All throughout their pregnancy period, the women were constantly transfused with plasma enriched with leukocytes from 16 donors. All the donors were erythrocyte compatible and the pregnancy in every woman was normal. All of them gave birth to a healthy baby (Fourth patient had a bicornuate uterus that was surgically repaired, but had two more miscarriages after the surgery).

The successful result of all the pregnancies can be explained by the existence of trophoblast/lymphocyte cross-reactive (TLX) antigens that stimulated the mother to produce a response with blastocyst protective factors and prevented maternal rejection in the antigenically special embryo.

The study found that the TLX antigens might be useful in organ transplantation. They may also help to prevent miscarriage. This was the first known use of lymphocyte immunization therapy in recurrent miscarriages [3].

7.5 Dr. Alan Beer

Dr. Beer suggested that paternal MHC antigen sharing could be linked to adverse pregnancy consequences. According to him, insufficient maternal recognition of paternal antigens can lead to low tolerance for the pregnancy. Therefore, it was suggested that immunizations with paternal mononuclear cells could improve maternal recognition of paternal alloantigens and allow patients with spontaneous, recurrent abortions (or other forms) to continue a pregnancy to term.

This was based on his study on the pregnancy outcomes of women with recurrent spontaneous abortions treated with LIT. Eleven women had abortions that took place after their immunization. This was compared with 14 women who gave birth to infants at term following LIT. The study found that the women who miscarried post-LIT had significant sharing of HLA A, B, and D/DR with their spouses and they failed to develop blocking antibodies even after LIT. They also had cytotoxic lymphocytes in the decidual lining and these cells were detrimental to pregnancy. It also suggested that in successful women, trophoblast leads to the migration of suppressive lymphocytes into the decidua and promotes the production of blocking antibodies specific to the paternal antigens [4].

7.6 Dr. Mowbray

Mowbray et al. conducted the first randomized, sequential controlled trial of LIT for spontaneous and unexplained abortions. The study included all couples who had three or more miscarriages without any cause and were not able to detect antipaternal antibodies.

Patients noticed a dramatic increase in success rates for women who had conceived quickly after treatment. Women who received purified lymphocytes from their husbands were more likely to have a successful next pregnancy than those who received their own lymphocytes. Results showed that 17 out of 22 women (78%) who were given paternal cells had successful pregnancies, while only 10 of 27 (37%) women had successful pregnancies when they were given their own cells.

7.6.1 Conclusions of This Study Were as Follows

- LIT is the treatment of choice for the group of selected women.
- Abortions seen in the women in the treated group can be due to other causes, suboptimal immunization or presence of anti-paternal cytotoxic antibodies and is only an indirect index of effective immunization [5].

7.7 Recurrent Miscarriages Immunotherapy Trialists Group (RMITG)

This was an international cooperative experimental study and analysis on Allogenic Leukocytes Immunotherapy for spontaneous abortions that are recurrent in nature. It was sponsored and organized by the ASRI (American Society for Reproductive Immunology) Ethics Committee (C.B. Coulam [USA], D.A. Clark [Canada], J.B. Smith [USA], J. Hill [USA], Alan Beer [USA)]).

A prospective meta-analysis and collaborative study that was conducted in 1994 confirmed the efficacy of LIT for women who had spontaneous abortions without anti-parental antibodies but not in those with anti-parental antibodies.

7.7.1 Aim

Conflicting results were reported from controlled clinical trials for allogeneic blood cell immunizations of women with RSA. This controversy was addressed by the analysis of raw data from all international patients who were enrolled in clinical trials that also included a control group. It was important to determine if alloimmune stimulation for the female partner increases the live birth rate. Because they had access to the appropriate raw data, 15 clinical centers worldwide were used for this purpose. Two separate data analysis teams independently evaluated nine randomized trials (7 double-blinded). This was to ensure robust conclusions. To test for bias in nonrandomized trials, one team also compared the results from six nonrandomized cohort-controlled trials. The factors that predict successful live births in couples with RSA were evaluated.

7.7.2 Results

The live birth ratios (ratios of live births between treatment and control groups) having 95% confidence intervals were (CI) 1.16 (CI 1.01–1.34; $P = 0.0331$) and 1.21 (CI 1.04–1.37; $P = 0.0224$). The results of randomized and nonrandomized trials were quite alike, despite the differences in the compositions of treatment and control groups. The live birth rate was lower for older female partners who had more than five abortions or had positive anticardiolipin antibodies. Live birth rates rose if the female partner had previously received treatment for serum antibodies to paternal lymphocytes, or was vaccinated (LIT) to convert from negative to positive. No evidence was found to show an increase in the risk of adverse reactions for the fetus.

7.7.3 Conclusion

These two independent global analyses of allogeneic Leukocyte Immunizations for Treatment of RSA revealed that alloimmunizations (LIT) can be an effective treatment for recurrent miscarriages. It is not possible to determine which patient subgroups would be most benefited from immunotherapy (LIT) as there is currently no test that can diagnose these patients. With more effective immunization protocols, the effectiveness of treatment could increase in this subgroup [6].

7.8 Daya S

This study was designed to analyze subgroups of data from a recently completed global collaborative meta-analysis based on raw data of patients with primary RSA that were enrolled in randomized controlled immunotherapy trials (LIT).

7.8.1 Study

This analysis included data from eight centers that had randomized controlled trials. Patients were included if they had suffered three or more spontaneous abortions, had no previous pregnancy past 20 weeks, had no identifiable reason for the abortions, and had no evidence of anti-paternal antibody.

7.8.2 Results

The center's meta-analysis showed that immunotherapy significantly increased the live birth rate. The common odds ratio being 1.94; 95% confidence interval (CI = 1.20–3.02). The chances of a successful outcome were much higher with treatment (relative risks = 1.46, 95% confidence interval (CI) 1.19–1.69). It took 16.3% to achieve an absolute treatment effect, meaning that six patients would need to be treated for one successful patient. For each additional pregnancy loss beyond three miscarriages, the likelihood of live birth was reduced by 23%.

7.8.3 Conclusion

When pretreatment anti-paternal antibody antibodies are absent, allogeneic leukocyte vaccination is an effective treatment. To identify patients who can benefit maximally from LIT, better diagnostic tests are needed [7].

7.9 Carp HJA et al. (1997)

7.9.1 Study

This study was published in Human Reproduction 1997 and focuses on women who have had five or more miscarriages. The analysis included both the data from patients who were treated in the past 8 years as well as the results of random and nonrandom trials that were reported by the Recurrent Miscarriage Immunotherapy Trialists Group. At the time of initial testing, there was no anti-paternal supplement-dependent antibody (APCA). The effect of paternal leucocyte vaccination on live birth rates was examined in primary, secondary, and tertiary aborters.

7.9.2 Results

(a) Primary aborters doubled the live birth rates with a benefit in excess of 31%.
(b) The live birth rate for Tertiary Aborters increased threefold with a 50% benefit.

7.9.3 Conclusion

LIT was shown to prevent miscarriage in primary/tertiary aborters with five or more miscarriages and who were APCA negative. To induce seroconversion it may be necessary to give booster immunization [8].

References

1. Medawar PB. Some immunological and endocrinological problems raised by the evolution of viviparity in vertebrates. Symp Soc Exp Biol. 1953;7:320–38.
2. Billingham RE, Brent L, Medawar PB. 'Actively acquired tolerance' of foreign cells. Nature. 1953;172:603–6.
3. Taylor C, Faulk WP. Prevention of recurrent abortion with leucocyte transfusions. Lancet. 1981;2(8237):68–70.
4. Beer AE, Semprini AE, Zhu XY, et al. Pregnancy outcome inhuman couples with recurrent spontaneous abortions: HLA antigen profile; HLA antigen sharing; Female serum MLR blocking factors; Paternal leucocyte immunization. Exp Clin Immunogenet. 1985;2:137–53.
5. Mowbray JF, Gibbins C, Liddel H, et al. Controlled trial of treatment of recurrent spontaneous abortion by immunization with paternal cells. Lancet. 1985;1(8435):941–3.
6. Recurrent Miscarriages Immunotherapy Trialists Group (RTIMG), Coulam CB, Clark DA, Collins J, Scott JR, Schlesselman JS, Aoki K, Carp HJA, Cauchi MN, Lim D, Christiansen OB, et al. Worldwide Collaborative Observational study and meta-analysison allogenic leucocyte immunotherapy for recurrent spontaneous abortion. Am J Reprod Immunol. 1994;32:55–72.
7. Daya S, Gunby J. The recurrent miscarriage trialist group. The effectiveness of allogenic leucocyte immunization in unexplained primary recurrent spontaneous abortions. Am J Reprod Immunol. 1994;32:294–302.
8. Carp HJA, Toder V, Torchinsky A, et al. Recurrent miscarriage immunotherapy trialists group. Allogenic leucocyte immunization after five or more recurrent abortions. Hum Reprod. 1997;12 (2):250–5.

Mechanism of Action of LIT

8

Abstract

LIT is an immunomodulatory treatment used for unexplained RPL/RIF. It is an active form of immunotherapy. LIT produces its effect through different immunological mechanisms, like production of blocking antibodies and activation of Treg cells leading to decrease in Th17/Treg ratio. It also acts through activation of Th2 type of cytokines (anti-inflammatory) leading to a decrease in Th1/Th2 cytokine ratio. These changes are made possible by the role of dendritic cells, which are found in the skin. These mechanisms result in immune tolerance, which prevents miscarriage and implantation failure.

Keywords

Mechanism of LIT · Blocking antibodies · Asymmetric antibodies · Treg cells · Th17/Treg ratio · Th1:Th2 balance

8.1 Introduction

Immunological response plays a very important role in the maintenance of pregnancy. Many theories have been put forward to explain the mechanisms that help to protect pregnancy (see Chap. 2). LIT's immunomodulatory effects have been explained by many immunological mechanisms. These mechanisms together or individually act on the feto–maternal interface to induce complex events preventing alloimmune rejection of pregnancy.

Alloimmune recognition plays a very important role in this by the formation of Blocking Antibodies. These antibodies have been measured in various assays and are known as cytotoxic antibodies, asymmetric antibodies or mixed lymphocyte reaction blocking antibodies [1–3]. Pregnant women who have experienced pregnancy loss are more likely to have low levels of asymmetric antibodies than their healthy equivalents [4].

M. Raut, M. Raut, *Lymphocyte Immunization Therapy (LIT) in Reproductive Failures*, https://doi.org/10.1007/978-981-19-2960-1_8

Numerous studies have shown a link between repeated abortions and parental similarity of Human Leukocyte Antigens (HLA). This causes hyporesponsiveness and prevents the production of blocking antibodies that act as immunological regulators to preserve the pregnancy.

These antibodies are known as Anti-paternal Cytotoxic Antibodies, Mixed Lymphocyte Reaction Blocking Antibodies, or Anti-idiotypic Antibodies that protect the fetus from the paternal HLA molecules. They also act as a barrier to the attack of the maternal T cells or NK cells [5].

8.2 The Mechanisms of Action of LIT (see Fig. 8.1) [6]

1. *Development of Blocking Antibodies*:

   Development of Blocking Antibodies (asymmetric antibodies) when presented with particulate or cellular antigen was shown by Margni [7].

   Various studies have confirmed the presence of BA and have been labeled differently. They are known as anti-paternal cytotoxic antibodies (APCAs) mixed lymphocyte reaction blocking antibodies (MLR-Babs or MLR-BF), or anti-idiotypic antibodies (Ab2).

   Mechanism

   Normally, antibodies have two Fab regions. After LIT, the antibodies that are produced have posttranslational high mannose glycosylations of one or both Fab chains. This makes the antibodies functionally univalent and is called asymmetrical antibodies. This asymmetric antibody is unable to form complexes, thus inhibiting complement fixation and phagocytosis. The other Fab fragment, being unmodified, has the antigen-binding capacity and it competitively inhibits the unmodified symmetric antibody.

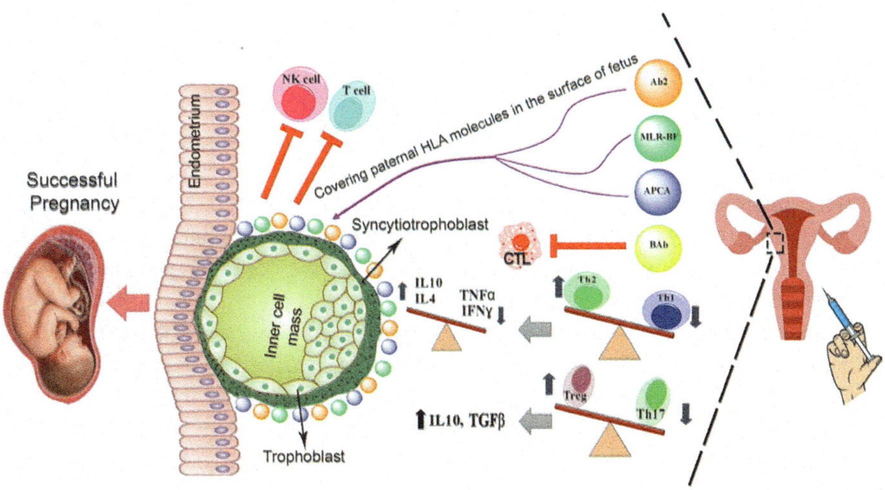

Fig. 8.1 Mechanisms of action of LIT [6]

These antibodies function as regulators and cover paternal HLA (human leukocyte antigen) molecules on the fetus and protect the fetus from the attack of maternal NK cells and T cells. The blocking antibodies might also prevent the cell-mediated immune response against fetal antigens.

Effect of LIT

LIT involves injecting paternal and/or donor lymphocytes into the dermis. These cells interact with macrophages, mast cells, and dendritic cells. Interaction with dendritic cells leads to allogeneic antigens being exposed to lymphocytes by the maternal immune system. This can lead to the formation of blocking antibodies that protect the pregnancy, as mentioned above [8].

Asymmetric antibody levels have also been shown to increase following LIT, as shown by Zenclussen et al. [9]

2. *Treg cell activation (Decrease in Th17/Treg ratio)*:
   T regulatory cells or Treg cells are suppressive CD4+ T cells with surface markers CD25+CD4+. They control immune responses to nonself-antigens and maintain natural self-tolerance.

   Although Treg cells recognize certain substances in the same manner as conventional immune cells, they can also turn off an immune reaction that might be initiated by conventional immune cells.

   Normal pregnancies have a greater number of T regulatory cells in their blood after ovulation. These cells seem to decrease in women who lose their pregnancies. In healthy pregnancies, however, the number of T regulatory cells keeps on rising upon the implantation of embryonic tissues into their endometrium. This protects the pregnancy from attack by conventional immune cells. The generation of tolerance-producing T regulatory cells decreases when embryonic tissues are also not recognized.

   LIT could increase tolerance by increasing the number and distribution of T regulatory cells. When a sufficient number of lymphocytes are injected into the dermis, they interact with dendritic cells, macrophages, and mast cells. This can lead to an increased number of Treg cells, which may prove beneficial in maintaining pregnancy. The Th17/Treg ratio was also significantly lower after immunotherapy. The continuation of pregnancy is possible by downregulating the Th17 population and upregulating the Treg population. This prevents miscarriage. Data also suggest that Th17/Treg balance regulation can be an important mechanism for treating unexplained recurrent loss of pregnancy [10].
3. *Effect on Th1:Th2 cytokine balance*:
   Serum secretion of TGF-β-1 and other Th2 cytokines is increased after LIT. This is accompanied by a decrease in the secretion of Th17-associated cytokine IL-17A as well as other Th1 cytokines. The Th1/Th2 ratio of cytokine is markedly decreased. This favors the continuation of pregnancy [11].
4. *Role of Dendritic cells*:
   Dendritic cells play an important role in the mechanism that protects pregnancy after LIT. Dendritic cells are the immunologic sensor cells present in the skin. These sensor cells interact with the antigen and present it to the immune system for a response. They then become migratory cells, which carry the antigen and

find their way to regional lymph nodes to deliver a tolerogenic signal. This results in further immunological changes that induce immune tolerance [12].

The maintenance of tolerance is achieved by dendritic cells turning on T regulatory cells that suppress immune responses to autologous tissue (preventing autoimmunity), semi-allogenic pregnancy (preventing miscarriage and implantation failure). It is unclear how dendritic cells make the decision to activate T regulatory cells or conventional immune cells. This is what has led to a debate about LIT's efficacy.

5. *Effect of LIT on HLA-G*:
   The key molecule in inducing tolerance to pregnancy is HLA-G. The trophoblastic cells express HLA-G molecules and not the classical HLA class I and class II genes. HLA-G can prevent the cytotoxic action of NK cells [13]. So, LIT can have a possible effect on mother's immune tolerance by regulation of HLA-G.

References

1. Pandey MK, Thakur S, Agrawal S. Lymphocyte immunotherapy and its probable mechanism in the maintenance of pregnancy in women with recurrent spontaneous abortion. Arch Gynecol Obstet. 2004;269:161–72.
2. Ito K, Tanaka T, Tsutsumi N, et al. Possible mechanisms of immunotherapy for maintaining pregnancy in recurrent spontaneous aborters: analysis of anti-idiopathic antibodies directed against autologous T-cell receptors. Hum Reprod. 1999;14:650–5.
3. Kishore R, Agarwal S, Halder A, et al. HLA sharing, anti-paternal cytotoxic antibodies and MLR blocking factors in women with recurrent spontaneous abortion. J Obstet Gynaecol Res. 1996;22:177–83.
4. Barrientos G, Fuchs D, Schrocksnadel K, et al. Low levels of serum asymmetric antibodies as a marker of threatened pregnancy. J Reprod Immunol. 2009;79:201–10.
5. Higuchi K, Aoki K, Kimbara T, Hosoi N, Yamamoto T, et al. Suppression of natural killer cell activity by monocytes following immunotherapy for recurrent spontaneous aborters. Am J Reprod Immunol. 1995;33:221–7.
6. Hajipour H, Nejabati HR, Latifi Z, Hamdi K, Bahrami-Asl Z, Fattahi A, Nouri. Lymphocytes immunotherapy for preserving pregnancy: mechanisms and challenges. Am J Reprod Immunol. 2018;80:e12853.
7. Margni RA, Perdigon G, Gentile T, et al. IgG precipitating and non-precipitating antibodies in rabbits repeatedly injected with soluble and particulate antigens. Vet Immunol Immunopathol. 1986;13:51–61.
8. Pandey MK, Agrawal S. Induction of MLR-Bf and protection of fetal loss: a current double blind trail of paternal lymphocyte immunization for women with recurrent spontaneous abortion. Int Immunopharmacol. 2004;4:289–98.
9. Zenclussen AC, Gentile T, Kortebani G, et al. Asymmetric antibodies and pregnancy. Am J Reprod Immunol. 2001;45:289–94.

10. Wu LI, Luo LH, Zhang YX, Li Q, Xu B, Zhou GX, Luan HB, Liu YS. Alteration of Th17 and Treg cells in patients with unexplained recurrent spontaneous abortion before and after lymphocyte immunization therapy. Reprod Biol Endocrinol. 2014;3(12):74.
11. Liu, et al. Low-dose lymphocyte immunotherapy rebalances the peripheral bloodTh1/Th2/Treg paradigm in patients with unexplained recurrent miscarriage. Reproduct Biol Endocrinol. 2017;15:95.
12. Steinman RM, Nussenzweig MC. Avoiding horror autotoxicus: the importance of dendritic cells in peripheral T cell tolerance. Proc Natl Acad Sci U S A. 2002;99(1):351–8.
13. Rizzo R, et al. The importance of HLA-G expression in embryos, trophoblast cells and embryonic stem cells. Cell Mol Life Sci. 2011;68:341–52.

9 Procedure of LIT

Abstract

LIT is a daycare procedure. There are different protocols of LIT in different centers. The couple is selected for LIT after immunological tests. Subsequent procedures including collection of blood, routes of administration, number of doses, dosages, the timing of procedure, indications, and contraindications are discussed in this chapter.

Keywords

LIT · Lymphocyte crossmatch · NK cells · K-562 test · CD57 · TNF-alpha · Treg · TH1:TH2 ratio · Procedure of LIT · Prerequisites for LIT

9.1 Introduction

Lymphocyte immunization therapy (LIT) is a daycare procedure. It essentially involves separation of lymphocytes from the blood of husband or donor and injecting them into the patient (wife). The protocols used for the LIT, done at different centers vary. The variation is with respect to quantity of blood taken, procedure of separation of lymphocytes, concentration of lymphocytes injected, routes of injection, number of doses, etc.

9.2 Indications

(a) Unexplained repeated pregnancy loss.
(b) Unexplained recurrent implantation failure.
(c) Unexplained infertility.

M. Raut, M. Raut, *Lymphocyte Immunization Therapy (LIT) in Reproductive Failures*, https://doi.org/10.1007/978-981-19-2960-1_9

9.3 Contraindications

(a) Known factors for RPL/RIF
(b) Selection criteria not fulfilled
(c) Presence of infection in husband (HIV, HCV, VDRL, HBsAg)—(donor lymphocytes can be used)

9.4 Criteria for Selection

Lymphocyte immunization therapy holds much potential as a safe and effective treatment approach for reproductive patients who have failed with less aggressive protocols when used correctly on a well-selected patient. Selection of a couple for LIT depends on many factors. The criteria for selection are as follows:

9.4.1 Group A: Clinical Criteria

1. History of Recurrent miscarriages: More than or equal to two miscarriages (*ASRM*).
2. Reports of all known causes of recurrent miscarriages are normal: Patients with chromosomal, anatomical, endocrine, and autoimmune causes of recurrent miscarriages should be excluded [1].
3. History of two or more unexplained IVF failures.
4. Unexplained infertility.

9.4.2 Group B: Laboratory Criteria (See Table 9.1)

Studies have suggested the need for laboratory criteria to identify couples who may benefit from LIT, as well as for the evaluation after LIT, before allowing the couple to attempt a new pregnancy. These tests are as follows:

9.4.2.1 Lymphocyte Crossmatch

(Synonyms: Leukocyte antibody detection—LAD, HLA crossmatch, Anti-paternal leucocyte antibody detection)

Introduction

LAD is a comprehensive test that helps to establish the cause of reproductive failure. It helps to find out whether there is an alloimmune cause of infertility or RM. It measures the amount of antibody that reacts with protein of non-self [paternal or donor] lymphocytes.

Table 9.1 Lab criteria—selection for LIT (Studies) [2]

Criteria	Related study
Higher levels of TNF-α (>400 pg/ml)	Gharesi-Fard et al.[36]
Negative for MLR-BAbs	Adachi et al.[64] Unander[108] Yang et al.[109]
Lack of chromosome abnormities history	Xu et al.[43]
Absence of pretreatment paternal antibodies	Daya et al.[116]
Presence of APCA after immunization	Chaichian et al.[113]
Absence of Antinuclear autoantibodies	Cavalcante et al.[119]
Absence of antithyroglobulin (TgAb) autoantibodies	Cavalcante et al.[119]
Number of previous miscarriages	Cowchock[123] Daya et al.[116]
Maternal and paternal age	Daya et al.[116] Kwak et al.[122]
CD4+CD25+ Treg cells levels	Not evaluated yet
Thl7 Level	Not evaluated yet
Lack of anatomical and hormonal defects	Not evaluated yet

TNF-α tumor necrosis factor-alpha, *MLR-Babs* mixed lymphocyte reaction blocking antibodies, *APCA* anti-paternal cytotoxic antibodies, *Th17* T helper 17, *Treg* regulatory T cell

Principle

Alloimmune recognition plays a very important role in this by the formation of Blocking Antibodies. These antibodies have been measured in various assays and are known as Antipaternal Cytotoxic Antibodies (APCA), Asymmetric antibodies or Mixed lymphocyte reaction blocking antibodies [3–5]. Women suffering pregnancy loss have significantly lower levels of asymmetric antibodies than their healthy counterparts [6].

Many studies confirm the association between recurrent abortions and parental similarity of Human Leukocyte Antigens (HLA) which induces hyporesponsiveness and inhibits the production of blocking antibodies as immunological regulators to maintain the pregnancy. Anti-paternal Cytotoxic Antibodies (APCAs), Anti-idiotypic Antibodies (Ab2), and Mixed Lymphocyte Reaction Blocking Antibodies (MLR-Bf) are considered as a part of regulators that cover paternal HLA molecules on the surface of fetus and make a barrier to prevent attack of the maternal T cells and NK cells [7].

Lymphocyte crossmatch or HLA crossmatch helps in the detection of anti-paternal antibodies in the mothers' blood.

Methods for Lymphocyte crossmatch

- Serology—Complement-Dependent Cytotoxicity method (CDC)—by Serology
- Flow Crossmatch (FCXM)—by flow cytometry
- Virtual Cross Match/Luminex XM—By Donor Specific Antibody (DSA) assay using Luminex platform.

Table 9.2 Results for lymphocyte crossmatch

	Results
Reaction of patient's serum (Neat) with donor's lymphocytes (HLA crossmatch)	Negative/Positive
Reaction of patient's serum (1:2) with donor's lymphocytes	Negative/Positive
Reaction of patient's serum (1:4) with donor's lymphocytes	Negative/Positive
Reaction of patient's serum with patient's lymphocytes (Autoantibodies)	Auto-antibodies detected/not detected

Table 9.3 Interpretation of the results

Interpretation	% Dead lymphocytes	[a]Scoring system
Negative	0–10	1+
Doubtful negative	11–20	2+
Weakly positive	21–40	4+
Positive	41–80	6+
Strongly positive	81–100	8+

[a]Scoring system: % cell death and interpretation is recorded as per ASHI guidelines

Results

CDC Crossmatch

This is the commonest method used to detect the presence of anti-paternal antibodies. This is done manually, using Complement Dependent Cytotoxicity method. It requires viable lymphocytes. In this method, the wife's serum is incubated with husband's lymphocytes. The results are interpreted as follows (see Tables 9.2 and 9.3):

Positive LAD patients can be further tested with Luminex anti-HLA testing to identify, which HLA antibodies are reacting with the husband's WBC. If there are no anti-paternal HLA antibodies, the test is labeled as false-positive test. This may be due to the presence of infection or other immune complexes in the blood.

Importance

The crossmatch between the serum of patient and the spouse's lymphocytes to detect anti-lymphocyte maternal antibodies is the most commonly used test to select a couple for the LIT procedure [8].

A negative lymphocyte crossmatch test in presence of unexplained RPL or RIF indicates an alloimmune cause and that immunomodulation (LIT) will be beneficial in such a case. The lymphocyte crossmatch may not be always positive in women who have a spontaneous successful pregnancy without a history of RPL or RIF.

It has been shown that the pregnancy outcomes of couples who had a positive crossmatch after immunotherapy are better [9, 10].

Mixed lymphocyte culture Blocking antibodies (MLC-Babs) can be used as a potential biomarker for indication and efficacy of paternal lymphocyte immunization in recurrent spontaneous abortion [11].

9.4.2.2 Assessment of the Peripheral NK Cells

Introduction

NK cells have a crucial role in the establishment of early pregnancy. They are present in blood and all tissues including the uterine endometrium. Abnormal function of NK cells can be responsible for RM or RIF.

Principle

Estimation of pNK cells can help to establish the possibility of alloimmune cause of RM or RIF. Also, pNK cell levels correspond to the levels of uNK cells.

Mechanism

pNK cells are a part of the innate immunity and are responsible for immune surveillance. They act on the cells, which are not allorecognized, killing them. They do not require activation.

Types

NK cells are a part of the innate immune system. They do not express the membrane molecules and the receptors seen on T cells and B cells. Human NK cells have been identified as $CD3^{-}CD56^{+}CD16^{+}$ or $CD3^{-}CD56^{+}CD16^{-}$. $CD56^{+}$ can be $CD56^{bright}$ or $CD56^{dim}$. The $CD56^{bright}$ are mainly in the peripheral blood. $CD56^{dim}$ are present in the endometrial lining. $CD56^{dim}$ cells can maturate into $CD57^{+}$ cells. The CD57+ NK cells are highly cytotoxic and are detrimental to pregnancy. They can be associated with RPL or RIF or Unexplained infertility [10].

Method

Flow cytometry.

Results

Different assays.

A. Proportion of pNK cells is measured out of all the lymphocytes.
   Proportion of pNK Cells: normal value—less than 18%.
B. Concentration.
   Concentration of activated CD56+dim cells. Determined with CD69 marker.
      Normal value: less than 12×10^6/l
C. In vitro assay NK cell inhibition.
   Inhibition of NK cell activity.

Interpretation

A. *Proportion of different lymphocytes* (see Table 9.4)

Normal range varies with each lab.
Normal levels:

CD3: 60–85%
CD19: 2.0–12.0%

Table 9.4 Natural killer cells

Test description	Observed value	Unit	Biological reference interval
Total Leucocytes (WBC) Count		/c.mm	4000–10,500
Lymphocytes		%	20–40
Absolute Lymphocyte Count		/c.mm	1000–3000
CD 3 Percentage		%	55–81
Absolute CD 3 Count		cells/μl	457–3926
CD (16+56) %		%	5.6–31
CD (16+56) Absolute cell count		/c.mm	173–1392

Population gated: Lymphocytes

CD56: 2.0–12.0%
CD19+ cells: 8.0–10.0%

NK cell inhibition panel.

IVIG and intralipid are used and inhibition of NK cells is tested at various dilutions.

Normal range varies with each lab.

Normal percentage
50:1 dilution - 10.0–40.0%
25:1 dilution - 5.0–30%
12.5:1 dilution - 3.0–20%
15–25% of women with RM have high NK cell levels [12].

B. *K-562 target cell test: (for NK cell activity)*

This test allows the quantitative determination of the cytotoxic activity of human natural killer (NK) cells. The K562 target cells are labelled with a lipophilic green fluorescent membrane dye in order to discriminate between effector and target cells. After the incubation period in the assay, killed target cells are identified by a DNA stain, which penetrates the dead cells and specifically stains their nuclei. In this way, the percentage of target cells killed by effector NK cells can be determined. The evaluation of NK activity is performed by flow cytometry [13].

Importance

Increased concentration or proportion of NK cells and abnormal NK cell inhibition assay indicate the presence of the alloimmune factor. These cases benefit from immunotherapy.

1. Meta-analysis of 22 studies done by Seshadri and Sankara showed that:
   (a) Increased pNK cells number in UI and RM patients
   (b) Increased pNK cells percentage in RM patients [14]
2. Analysis of peripheral blood lymphocytes in 76 women with recurrent miscarriage before and after lymphocyte immunization therapy showed that the percentages of CD56(+) CD3(−) NK cells and IFN-γ-producing cells were significantly lower after LIT in the successful pregnant group compared to those in the abortion group, indicating that LIT alters the proportions and functions of most peripheral blood lymphocyte subsets which may be beneficial for pregnancy maintenance [15].
3. The patients with unexplained RPL show higher proportions of CD3+DR+ T lymphocytes with higher proportions and absolute numbers of CD4+DR+ and CD8+DR+ T lymphocytes and lower CD16+CD56+ NK-cells [16].
4. The percentage of CD56+ NK cells in peripheral blood in normal healthy individuals varies from 5% to 29% [17]. However, a finding of more than 12% NK cells in women with infertility or miscarriage has been defined as abnormally raised and used as an indication for treatment [18].
5. It has been shown that population-based studies to ascertain normal levels of NK cells have included men (known to have significantly higher levels than women) and not excluded women with reproductive failure. So, it gives the significant level of NK cells in RPL/RIF is above 18%. In RIF, 13% of women have elevated concentrations of NK cells (CD56dim CD69+), indicating that NK cell abnormalities indicate an immune implantation disorder [19].

9.4.2.3 Assessment of Endometrial NK Cells

Introduction

NK cells in the uterine lining or endometrium play an important role in the implantation and maintenance of early pregnancy. Excess concentration and activity of uNK cells have been shown to be associated with RM and RIF.

Origin of Cells

uNK cells are a part of innate immunity. Majority of uNK cells are recruited from the peripheral blood NK cells. Their concentration is 10% in the proliferative phase, 20% in the late secretory phase, and more than 30% in early pregnancy.

Human NK cells do not express the membrane molecules and the receptors seen in T cells and B cells. They have been identified as $CD3^{-}CD56^{+}CD16^{+}$ or $CD^{-}CD56^{+}CD16^{-}$. $CD56^{+}$ can be $CD56^{bright}$ or $CD56^{dim}$. The $CD56^{bright}$ are mainly in the peripheral blood. $CD56^{dim}$ are present in the endometrial lining. $CD56^{dim}$ cells in the endometrial lining can maturate into $CD57^{+}$ cells. The CD57+ NK cells are highly cytotoxic and are detrimental to pregnancy. They can cause RPL or RIF [20].

Mechanisms

uNK cells interact with trophoblast cells and may regulate the physiological invasion of trophoblast by allorecognition. Overactivity of uNK cells can cause RM or IF by direct cytotoxicity of cytokine production.

Principle

Estimation of uNK cells can help to establish the possibility of alloimmune cause for RM or IF.

Methods

uNK cells are assessed by taking an endometrial biopsy and testing it for the presence of uNK cells and its concentrations.

Timing

EB is done between 19th and 23rd day of the cycle.

Collection of Sample

The endometrial biopsy material is collected in formalin (60%), that is diluted by 50% (1:1 with water).

Method of Testing

A. Immunohistochemistry
   Easier method, can only count CD56 and CD57 cells.
   Measurement of subtypes or levels of activation is not possible.
   Reliable and consistent method of counting is required.
B. Flow Cytometry
   Technically difficult
   Can assess the surface makers CD56+ bright/CD56+ dim.

Results

Positive: CD57 cells present.
Negative: Absence of CD57 cells.
15–25% of RM show high uNK cells.

Interpretation

This test has limitations because endometrial tissue is complex with cell clusters and uNK cell counts may vary to a great extent, and the uNK cell numbers change rapidly daily in the luteal phase.

Importance

Increased uNK cell CD56 or presence of CD57 indicates a higher association with alloimmune cause for RIM or RIF.

Future Prospects

uNK cell estimation is a highly specialized investigation. A dedicated and interested pathologist is required for the accuracy of results. Daily reference ranges for a given population are required to be prepared. This will help in better interpretation of results.

9.4.2.4 Serum TNF-α

Introduction

Tumor Necrosis Factor (TNF-α) is a cytokine involved in systemic inflammation. TNF-alpha plays an important role in the maintenance of pregnancy and in alloimmune rejection leading to RM or RIF.

Principle

TNF-α is a Th1 type of cytokine. The different cells in endometrium (lymphocytes including NK cells, fibroblasts, and other immune cells) express TNF-α and is required for a variety of normal endometrial function. It is pro-inflammatory and elevated levels are associated with RM or RIF.

Mechanism

TNF-α is produced by the activated macrophages and NK cells. Elevated levels have deleterious effects on oocyte, implantation of embryo, and future growth of embryo by inducing inflammatory reaction. Along with other pro-inflammatory cytokines like IFN-γ, it causes apoptosis of corpus luteal cells causing luteal insufficiency and subsequent pregnancy loss.

Method

Plain blood was collected and serum separated. Done by chemiluminescent assay.

Results and Interpretation

Normal level: 4–12 ng/ml.

Raised levels of TNF-α with a history of uRPL/uRIF indicate alloimmune problem and immunomodulatory treatment (LIT) will be beneficial. In the studies conducted by Daher S. et al. and Kim H. R. et al., significantly higher levels of TNF-α was observed in patients with RPL compared to controls [21, 22].

Importance

Elevated levels of serum TNF-α are associated with RM or IF or IVF failure. Even after treatment with immunotherapy, the success rates are limited in the presence of elevated serum TNF-α. Levels of TNF-α should be brought to normal with medications before planning pregnancy or IVF cycle.

Future Prospects

Use of anti-TNF-α medicines help in normalizing the level of TNF-α. These medicines also lead to immunosuppression, making the patient prone to other disorders. Better anti-TNF-α medicines in future will definitely help to overcome this problem, which can be treated effectively.

9.4.2.5 Assessment of Treg Cells

T-Regulatory cells, Treg, are the subset of T lymphocytes with potent suppressive activity and are important in curtailing destructive immune responses and preventing autoimmune disease.

Apart from Th1/Th2 responses, Th17 and Treg cell responses are also important. T17 cells produce pro-inflammatory cytokine IL-17, leading to inflammatory response. In contrast, Treg cells play an important role in immunoregulation and induction of tolerance. Treg cells inhibit proliferation and cytokine production in CD4+ and CD8+ T cells.

Immunoglobulin production by these cells, decrease in cytotoxic activity of natural killer NK cells, and maturation of dendritic cells (DC) result in induction of tolerance [20].

Good levels of Treg cells are essential to prevent production of autoantibodies (preventing autoimmune disease). They are important for reproductive immune tolerance to allow the embryos to implant (preventing IVF failure), and reduce the risk of miscarriage.

Testing

Treg (CD4 CD25 fox P3+)

Normal limit: more than or equal to 0.7%.

Importance

It has been shown by Wu et al. that in patients with URSA, immunotherapy (LIT) with paternal mononuclear cells leads to the secretion of the Treg-associated cytokine TGF-β, and Th2 cytokines, while the secretion of Th17-associated cytokine IL-17A and Th1 cytokine production was decreased. The Th1/Th2 cytokine ratio significantly decreased. The Th17/Treg ratio significantly decreased. This Th2 and Treg bias was beneficial for pregnancy [23].

9.4.2.6 Assessment of Th1:Th2 Cytokine Ratio

Introduction

This is a test to determine the balance between the Th1 [pro-inflammatory] and Th2 [anti-inflammatory] immune response and its possible role in repeated miscarriages or implantation failure.

Principle

Th1 cells are part of cell-mediated immunity and release various cytokines in response to foreign proteins. It produces a pro-inflammatory response. Th2 cells mediate antibody-mediated response and it is anti-inflammatory. The test determines which response is dominant.

Mechanism

Th1 cells produce cytokines interferon-γ, IL-1B, TNF-α leading to inflammatory response. Th2 cells produce cytokines IL4, IL-5, IL-6, IL-10, and IL-13 leading to anti-inflammatory response.

Table 9.5 Th1:Th2 balance

Cytokines	TH1 dominance	TH2 dominance
IL2	High	Low
IL4	Low	High
IL12	High	Low
IFN gamma	High	Low
TNF alpha	High	Low

Method
Luminex processing of Th1 cytokines—[IL-2, IFN-γ, TNF-α, Th2 cytokines] IL4, IL5, IL10]. Fasting is not required. T-helper cells are stimulated so that they produce cytokines that indicate whether they are Th1- or Th2-type cells. The cells are then reacted with antibodies that detect these cytokines. The cells of each type are then counted by flow cytometry.

Results and Interpretation (See Table 9.5)
There can be three states:

A. Balanced
B. Strong Th1 dominance
C. Strong Th2 dominance

Cytokine panel level 1: IL1, IL6, IL10, TNF-α.

Cytokine panel level 2: IL1, IL2, IL4, IL5, IL6, IL10, IL13, IFN-γ, TNF-α, TNF-β, TGF-β.

Th1:Th2 cytokine ratios can also be studied:

Normal limits
TNF-α:IL10 - 13.2–30.3
IFN-γ:IL10 - 5.6–21.0

Raised ratios indicate inflammatory response

Importance
Th1 dominance indicates alloimmune factor responsible for repeated miscarriages or implantation failure.

Kheshtchin N et al. showed that the ratios of Th1/Th2 cytokine receptors were higher in women with RPL compared to controls. The ratio of Th1/Th2 cytokine receptors was decreased in women with RM after LIT, while no significant change was identified in the ratio after immunotherapy. The study indicated that LIT might influence pregnancy outcome via a shift in the balance of the Th1/Th2 cytokine receptors [24].

9.4.2.7 Conclusion for Investigations

The selection of couples, likely to be benefited from LIT, will become more efficient with the development of more effective tests and considering the evidence of clinical and other immunological abnormalities. For other tests that may be used in diagnosis, please see Chap. 5.

9.5 Prerequisites to LIT

1. Immunological testing:
   This is done to select the couple likely to be benefited from LIT.
2. Infection profile of both the partners or the patient and the donor (in case of use of donor lymphocytes).
   These tests include HIV, HCV, VDRL, and HBsAg.
3. Blood group:
   Blood group of both the partners (or donor) to know about Rh status. If the wife is Rh-negative and husband/donor is Rh-positive, a prior Rh antibody titer should be done. In such patients, post-LIT, the patient is given Inj. Anti D (300 mcg) to prevent Rh iso-immunization.
4. Sr. IgA:
   This is done if there is a history of allergy in the patient.

9.6 Procedure

LIT involves the separation of paternal lymphocytes, and injecting them into the wife. With respect to the procedure, the following factors are important and must be considered before giving the injections:

1. Route of administration
   During immunization with paternal lymphocytes, it is important to inject the cells intradermally so that there is adequate physical interaction with dendritic cells. Sometimes, the cells are injected by subcutaneous and intravenous routes also [25].
   Yu et al. investigated various methods of administration of lymphocytes and demonstrated that intradermal immunization gives the best results [8].
2. Concentration of cells:
   In vaccination, very low and high doses normally induce tolerance whereas intermediate doses induce immunization. Liu et al. also observed that the results were better when the concentration of lymphocytes used in the LIT was lower than100 $\times$ 10^6/ml per application, when compared with a concentration greater than 100 $\times$ 10^6/ml per application [26].
3. Temperature of lymphocyte preparation:
   Initially, the lymphocytes were prepared at 4 °C (Ober). However, Clark and Chaouat [27] have shown that if stored at 4 °C, there is a loss of CD200 expression which is important in the development of immunotolerance. Pandey has reported a higher rate of successful live births when the cells were prepared at 37 °C before use [28]. The lymphocytes should not be kept at room temperature for a long time.

4. Number of doses:
   The protocol related to doses of lymphocyte injection varies in different centers. Inoculation is repeated till a sufficient immunological response is developed. Many centers use two doses two weeks apart while some use more, till sufficient immunological response is seen. Some used to give four intradermal doses every 2–3 weeks [29].
5. Time allowed for immunologic response to develop:
   It takes time for the optimal immunological response to develop after LIT. Usually, a period of 4 weeks is considered adequate for the development of maximum protective response [3].
6. Timing with respect to the menstrual cycle: Not related.
7. Time interval between paternal blood collection (lymphocyte separation) and Lymphocyte injection:
   The lymphocyte preparation should be injected as soon as it is ready for optimum results. In the Ober study, the cell preparation was stored overnight and injected the next day. This reduced the immunogenicity leading to inferior results. Storage of lymphocytes may alter their immunogenetic properties radically, which may impair the anti-abortive effect (see Chap. 10).
8. Booster doses:
   Usually, lymphocyte injections are given before pregnancy. The couple is asked to make attempts at conception, once there is sufficient evidence of immunological response. Sometimes, the booster dose is given during early pregnancy. When immunization with lymphocytes was performed before and during pregnancy, the results were better (Liu et al.) [26].

   The mechanics of LIT are relatively simple. Its consistent and successful practice calls for an understanding of the immune system in healthy pregnancy, as well as the immune response in the skin upon exposure to injected antigen. As we learn more about the changes occurring at cellular and molecular levels, it will be easier for us to understand Lymphocyte Immunization Therapy in a better manner.

References

1. Coulam CB, Stephenson M, Stern JJ, et al. Immunotherapy for recurrent pregnancy loss: analysis of results from clinical trials. Am J Reprod Immunol. 1996;35:352–9.
2. Hajipour H, Nejabati HR, Latifi Z, et al. Lymphocytes immunotherapy for preserving pregnancy: mechanisms and challenges. Am J Reprod Immunol. 2018:e12853. https://doi.org/10.1111/aji.12853.
3. Pandey MK, Thakur S, Agrawal S. Lymphocyte immunotherapy and its probable mechanism in the maintenance of pregnancy in women with recurrent spontaneous abortion. Arch Gynecol Obstet. 2004;269:161–72.
4. Ito K, Tanaka T, Tsutsumi N, et al. Possible mechanisms of immunotherapy for maintaining pregnancy in recurrent spontaneous aborters: analysis of anti-idiopathic antibodies directed against autologous T-cell receptors. Hum Reprod. 1999;14:650–5.

5. Kishore R, Agarwal S, Halder A, et al. HLA sharing, anti-paternal cytotoxic antibodies and MLR blocking factors in women with recurrent spontaneous abortion. J Obstet Gynaecol Res. 1996;22:177–83.
6. Barrientos G, Fuchs D, Schrocksnadel K, et al. Low levels of serum asymmetric antibodies as a marker of threatened pregnancy. J Reprod Immunol. 2009;79:201–10.
7. Higuchi K, Aoki K, Kimbara T, Hosoi N, Yamamoto T, et al. Suppression of natural killer cell activity by monocytes following immunotherapy for recurrent spontaneous aborters. Am J Reprod Immunol. 1995;33:221–7.
8. Yu HL, Deng XH, Chao L, Chen C, Han YL. Study on positive rate of blocking antibody in women with recurrent spontaneous abortion administered by route and frequency of paternal lymphocyte immunotherapy. Zhonghua Fu Chan Ke ZaZhi. 2013;48:903–6.
9. Coulam CB, Clark DA, Collins J, Scott JR. Worldwide collaborative observational study and meta-analysis on allogenic leukocyte immunotherapy for recurrent spontaneous abortion. Am J Reprod Immunol. 1994;32:55–72.
10. Nielsen CM, White MJ, Goodier MR, Riley EM. Functional significance of CD57 expression on human NK cells and relevance to disease. Front Immunol. 2013;4:422. https://doi.org/10.3389/fimmu.2013.00422.
11. Khonina NA, Broitman EV, Shevela EY, Pasman NM, Chernykh ER. Mixed lymphocyte reaction blocking factors (MLR-Bf) as potential biomarker for indication and efficacy of paternal lymphocyte immunization in recurrent spontaneous abortion. Arch Gynecol Obstet. 2013;288:933–7.
12. Sacks G. NK cells in peripheral blood and the endometrium. In: Christiansen OB, editor. Recurrent pregnancy loss. West Sussex: Wiley; 2014.
13. Kandarian F, Sunga GM, Arango-Saenz D, Rossetti M. A flow cytometry-based cytotoxicity assay for the assessment of human NK cell activity. J Vis Exp. 2017;126:56191.
14. Seshadri S, Sunkara SK. Natural killer cells in female infertility and recurrent miscarriage: a systematic review and meta-analysis. Hum Reprod Update. 2014;20(3):429–38. https://doi.org/10.1093/humupd/dmt056.
15. Liang P, Mo M, Li GG, Yin B, Cai J, Wu T, et al. Com-prehensive analysis of peripheral blood lymphocytes in 76women with recurrent miscarriage before and after lymphocyte immunotherapy. Am J Reprod Immunol. 2012;68:164–74.
16. Kuona RJ, Schaumanna J, Goeggl T, Strowitzki T, Sadeghi M, Opelz G, Daniel V, Totha B. Patients with idiopathic recurrent miscarriage show higher levels of DR+ activated T-cells that are less responsive to mitogens. J Reprod Immunol. 2015;112:82–7.
17. Bisset LR, Lung TL, Kaelin M, Ludwig E, Dubs RW. Reference values for peripheral blood lymphocyte phenotypes applicable to the healthy adult population in Switzerland. Eur J Haematol. 2004;72:203–12.
18. Kwak JY, Kwak FM, Gilman-Sachs A, Beaman KD, Cho DD, Beer AE, et al. Immunoglobulin G infusion treatment for women with recurrent spontaneous abortions and elevated CD56+ natural killer cells. Early Preg. 2000;4:154–64.
19. Sacks G, Yang Y, Gowen E, Smith S, Fay L, Chapman M. Detailed analysis of peripheral blood natural killer cells in women with repeated IVF failure. Am J Reprod Immunol. 2012;67:434–42.
20. Nielsen CM, White MJ, Goodler MR, Riley EM. Functional significance of CD57 expression on human NK cells and relevance to disease. Front Immunol. 2013;4:422.
21. Daher S, Camano L. Cytokines in recurrent pregnancy loss. J Reprod Immunol. 2004;62(1-2):151–7.
22. Kim HR, Park AJ, Lee MK. CD56/CD16 expression on mononuclear cells and concentration of serum TNF-alpha in recurrent spontaneous abortion. Korean J Lab Med. 2006;26(3):198–203.
23. Wu L, Luo LH, Zhang YX, Li Q, Xu B, Zhou GX, et al. Alteration of Th17 and Treg cells in patients with unexplained recurrent spontaneous abortion before and after lymphocyte immunization therapy. Reprod Biol Endocrinol. 2014;12:74.

24. Kheshtchin N, Gharagozloo M, Andalib A, Ghahiri A, MaracyMR RA. The expression of Th1- and Th2-related chemokine receptors in women with recurrent miscarriage: the impact of lymphocyte immunotherapy. Am J Reprod Immunol. 2010;64:104–12.
25. Kwak JY, Gilman-Sachs A, Beaman KD, et al. Reproductive outcome in women with recurrent spontaneous abortions of alloimmune and autoimmune causes: preconception versus postconception treatment. Am J Obstet Gynecol. 1992;166:1787–95.
26. Liu Z, Xu H, Kang X, Wang T, He L, Zhao A. Allogenic lymphocyte immunotherapy for unexplained recurrent spontaneous abortion: a meta-analysis. Am J Reprod Immunol. 2016;76: 443–53.
27. Clark DA, Chaouat G. Loss of surface CD200 on stored allogeneic leukocytes may impair anti-abortive effect in vivo. Am J Reprod Immunol. 2005;53:13–20. https://doi.org/10.1111/j.1600-0897.2005.00240.x.
28. Pandey MK, Agrawal S. Induction of MLR-Bf and protection of fetal loss: a current double blind randomized trial of paternal lymphocyte immunization for women with recurrent spontaneous abortion. Int Immunopharmacol. 2004;4(2):289–98. https://doi.org/10.1016/j.intimp.2004.01.001.
29. Takeshita T. Diagnosis and treatment of recurrent miscarriage associated with immunologic disorders: is paternal lymphocyte immunization a relic of the past? J Nippon Med Sch. 2004;71 (5):308.

Safety of LIT

10

Abstract

For over three decades, Lymphocyte Immunization Therapy (LIT) has been used in uRPL/uRIF. LIT can have untoward side effects, which has always concerned clinicians as well as immunologists. Numerous studies have examined the side effects and potential complications of LIT. Although some studies suggested an increased risk of infection, most studies have not found any significant complications. The large study of Kling et al. that analyzed the adverse effects of LIT in more than 3000 patients showed no acute allergic reactions (anaphylaxis) or increased incidence of autoimmunity or graft versus host disease. Local side effects in the form of bluish-reddish spots and skin rashes were seen in around 8% of patients. LIT is therefore a low-risk and safe procedure.

Keywords

Side effects of LIT · Anaphylaxis · Graft versus host disease (GvHD) · Skin rashes · Autoimmunity · Effect of LIT on baby

10.1 Introduction

Current evidence supports the safety of Lymphocyte Immunization Therapy LIT [1]. It is considered a low-risk procedure [2].

10.2 Effects of LIT

Before the endometrium can become a receptive surface for embryo in a healthy, natural pregnancy, there is a period of tolerogenic immune response to paternal antibodies. Their development takes time, as with all adaptive immune responses. LIT can take many weeks to develop the optimal response. LIT done just before

M. Raut, M. Raut, *Lymphocyte Immunization Therapy (LIT) in Reproductive Failures*, https://doi.org/10.1007/978-981-19-2960-1_10

conception might not allow for sufficient time to develop the maximum protective response. For a sufficient immune response to develop, it takes 4–6 weeks.

10.3 Side Effects

As LIT involves separation and injection of blood components (lymphocytes), ensuring complete safety is important.

1. The first data related to safety of LIT was published in 1994 (Coulam CB et al.). The study showed that 2.1% of women treated with lymphocytes showed complications, compared to 0.5% in the control group. The common maternal complications were Viral infections (CMV, Hepatitis), flu-like symptoms, and fever. There was no increased incidence of fetal complications like Preterm birth, IUGR, Fetal death, and neonatal complications like thrombocytopenia and congenital malformations in the patients treated with LIT, compared to the control [3].
2. Kling et al. (2006) [4]:
   This survey evaluates all possible side effects of LIT. It focuses on both the immediate and the delayed systemic effects of LIT. The survey had two parts, first was the prospective 4-week follow-up of patients (2000–2003) after the LIT and second was retrospective feedback 2–3 years after the LIT (1996–2002). The patients studied were with both IVF failure and recurrent miscarriages.

   The side effects of LIT can be divided into two groups:

1. Early—Local or systemic
2. Delayed—Local or systemic

10.3.1 Early Side Effects

1. Early local side effects
   These are seen during the first week after the LIT and usually subside in 3 weeks.
   (a) *Bluish mark*
   A faint, bluish mark can be seen at the site of intradermal injections. It usually disappears within 2–3 weeks without swelling or scarring.
   (b) *Blister formation*
   This is seen at the intradermal injection site in about 6% of patients. It subsides spontaneously without suppuration or any other complications. It is more commonly seen when the lymphocyte concentration is high.
   (c) *Skin rash and urticaria*
   This is seen in less than 1% of patients after LIT. The rashes are transient and usually subside spontaneously. Antihistaminics or short-term steroids are required rarely.

(d) *Other side effects*

There are other mild side effects like redness, itching, burning, and swelling at the intradermal injection sites. Rarely, there can be axillary lymphadenopathy.

2. Early systemic side effects:

(a) *General symptoms*

Mild fever, fatigue, headache, and dizziness may be seen in some patients. Rarely, there can be nausea, vomiting or diarrhea.

(b) *Specific systemic complaints*

Commonest specific systemic complaint is skin rashes. The rashes are very often urticarial and associated with itching. They are usually transient and very rarely recur on and off for a few months. There is no systemic effect on organs like liver, kidney, or lungs.

(c) *Anaphylactic reaction*

Anaphylactic reaction has not been reported in the literature related to LIT.

3. Significance of early side effects

Side effects that occur early are usually mild and resolve quickly. These side effects are caused by the immune system's reaction to lymphocyte antigens. They are not due to Graft versus Host Disease (GvHD).

10.3.2 Delayed Side Effects

1. *Delayed local side effects*

There are no delayed local side effects. There is no pigmentation, granuloma, or scarring.

2. *Delayed systemic side effects*

(a) Development of Autoimmune disorder

It was thought that LIT might predispose the patient to autoimmune disease. This was considered because LIT involves injection of allogenic cells into the patient, creating microchimerism. In the case of microchimerism, there is persistence of allogeneic cells in the host in a stable asymptomatic equilibrium. However, studies have proved now that microchimerism does not predispose to autoimmunity.

Retrospective meta-analysis showed that LIT had a lower incidence of autoimmune diseases than the control group [5].

Even, the incidence of autoimmune disease in the Kling study was 0.2%, which corresponds to the incidence expected in the female population with infertility, which is more than or equal to 0.1%.

(b) *Graft versus Host Disease*

Allogeneic viable T cells can cause Graft versus Host Disease, which can enter circulation. Studies have shown that LIT does not predispose to GvHD. It has been shown that GvHD may result if the allogeneic cells are injected into the circulation. While, in LIT, the injections are given mainly intradermal and sometimes, subcutaneous also. The high concentration of dendritic cells in the skin helps to generate a good immune response in LIT and prevents GvHD.

(c) *Transmission of Infections*
There is a potential risk of transmission of infections from husband to wife. However, this fear is far from true. With advanced laboratory methods for detection of infections, there is no risk of infection transmission associated with LIT. Furthermore, in the case of LIT, the person donating the lymphocytes and the one receiving it are most of the time married. This means that they have already been exposed to infections that can be potentially transmitted.

10.4 Effect of LIT on Infants

Many studies have shown that LIT in a mother does not affect the health of the infant [6, 7]. The incidence of congenital anomalies was not significantly different in the immunized and the nonimmunized mothers [8]. No appreciable difference in the rate of low birth weight babies and other complications has been seen either [9].

10.5 Conclusion

(a) LIT's early side effects are mild for immunocompetent women who are in the childbearing age group.
(b) Intradermal LIT does not induce anaphylaxis, autoimmunity or graft against host disease, as per the evidence.
(c) There is no risk of transmission of infections.

LIT is, therefore, in general, a safe procedure that does not cause any complications or side effects.

References

1. Cavalcante MB, et al. Lymphocyte immunotherapy in the treatment of recurrent miscarriage: systematic review and meta-analysis. Arch Gynaecol Obstet. 2017;295:511–8. https://doi.org/10.1007/s00404-016-4270-z.
2. Gunther V, et al. Active immunisation with partner lymphocytes in female patients who want to become pregnant–current status. Geburtsh Frauenheilk. 2018;78:260–73.
3. Coulam CB, Clark DA, Collins J, Scott JR. Worldwide collaborative observational study and meta-analysis on allogenic leukocyte immunotherapy for recurrent spontaneous abortion. Am J Reprod Immunol. 1994;32:55–72.
4. Kling C, Steinmann J, Westphal E, Magez J, Kabelitz D. Adverse effects of intradermal allogeneic lymphocyte immunotherapy: acute reactions and role of autoimmunity. Hum Reprod. 2006;21(2):429–35. https://doi.org/10.1093/humrep/dei316.

5. Coulam CB, Clark DA, Collins J, Scott JR, Schlesselman JS, Aoki K, Carp HJA, Cauchi MN, Lim D, Christiansen OB, et al. Worldwide collaborative observational study and meta-analysis on allogenic leucocyte immunotherapy for recurrent spontaneous abortion. Recurrent Miscarriages Immunotherapy Trialists Group (RTIMG). Am J Reprod Immunol. 1994;32:55–72.
6. Christiansen OB, Mathiesen O, Husth M, Lauritsen JG, Grunnet N. Placebo-controlled trial of active immunization with third party leukocytes in recurrent miscarriage. Acta Obstet Gynecol Scand. 1994;73(3):261–8. https://doi.org/10.3109/00016349409023451.
7. Cavalcante MB, da Silva Costa F, Júnior EA, Barini R. Risk factors associated with a new pregnancy loss and perinatal outcomes in cases of recurrent miscarriage treated with lymphocyte immunotherapy. J Matern Fetal Neonatal Med. 2015;28(9):1082–6. https://doi.org/10.3109/14767058.2014.943175.
8. Carp HJ, Ben-Shlomo I, Toder V, Nebel L, Mashiach S. Congenital malformations after immunotherapy for habitual abortion: is there an increase? Gynecol Obstet Investig. 1993;36(4):198–201. https://doi.org/10.1159/000292629.
9. Nonaka C, et al. Studies on the infants born to the patients with Unexplained Recurrent Abortion who underwent immunotherapy using their husband's lymphocytes. Austin J Obstet Gynecol. 2017;4(1):1067.

Controversies About LIT: REMIS Study 11

Abstract

Lymphocyte Immunization Therapy was a commonly recommended treatment for patients with unexplained RPL in the nineties. The REMIS study, a large placebo-controlled trial from the USA, was published in Lancet and it showed that LIT is not beneficial in unexplained RPL. It discouraged the use of LIT. However, subsequent research pointed out flaws in REMIS and proved that the REMIS findings were not correct. The study had flaws in patient selection, lymphocyte preparation, other methodologies, and follow-up. However, it had an impact on the recommendations of Cochrane reviews, which gave recommendations against the use of LIT in unexplained RPL.

Keywords

REMIS study · Ober study · CD200 · Criticism of REMIS study

11.1 Controversies About LIT

LIT was accepted as a common treatment and became very popular until 1999 when results from a large placebo-controlled study revealed that LIT did indeed decrease the chance of live births compared to placebo (REMIS Study) [1].

11.2 REMIS Study

In the early 1990s, immunization with paternal mononuclear cells was used as a method to treat unexplained recurrent abortions. It was believed that it was due to alloimmune rejection or a genetically foreign pregnancy. However, results from different studies were inconsistent and confusing. A randomized controlled trial was therefore conducted in six US centers between 1992 and 1997. This was the

M. Raut, M. Raut, *Lymphocyte Immunization Therapy (LIT) in Reproductive Failures*, https://doi.org/10.1007/978-981-19-2960-1_11

Recurrent Miscarriage Investigation Study (REMIS). The REMIS study examined whether immunization with paternal mononuclear cells improves the success rate of pregnancies.

11.3 Observations of REMIS Study (See Fig. 11.1) [1]

- Women who had recurrent miscarriages did not experience any improvement in their pregnancy outcomes after being immunized with paternal mononuclear cell vaccines.
- Nearly 65% of the control patients who were pregnant became successful, despite having a history of unexplained repeated miscarriages. This was compared to 46% of those in the treated group.
- The immunized group who became pregnant had a higher miscarriage rate and a lower gestational age.
- REMIS recommended against LIT as an option for unexplained, recurring miscarriage as they did not see any benefits.

11.4 Criticism of REMIS Study

Lancet published REMIS study in 1999. Recurrent miscarriage women did not benefit from immunization with paternal mononuclear cells. Due to the lack of benefit, the REMIS study recommended against LIT as a treatment for unexplained recurrent miscarriages.

Many irregularities were found in the REMIS study. The most critical points were:

Pregnancy outcome	Treatment group	Control group
All randomised patients		
Total	86	85
Success	31 (36%)	41 (48%)
Failure	55 (64%)	44 (52%)
All pregnant patients		
Total	68	63
Success	31 (46%)	41 (65%)
Failure	37 (54%)	22 (35%)

Table 2: **Study outcomes by treatment group**

THE LANCET • Vol 354 • July 31, 1999

Fig. 11.1 Results of REMIS study [1]

1. The REMIS is the only study, till date, that has shown a negative effect of immunotherapy with lymphocytes on live births.
2. *Storage of blood before separation of lymphocytes:* The concentrate of paternal lymphocytes was prepared by Ober et al. with blood that had been stored for several hours at a temperature of 1–6 °C. This extended the interval between the collection of the blood of the spouse and the immunization [1]. This destroyed the immunomodulatory effect of the mononuclear cells.
   Later Clark et al. demonstrated that the lymphocytes have a surface marker, CD200, which is important for the immunomodulatory effect of lymphocytes. Hence, an adequate number of CD200-positive lymphocytes are required in the cell preparation, for LIT to become effective. It was also proven that keeping total blood at low temperatures can decrease the count of CD200-positive cells, leading to a loss of immunomodulatory activity of lymphocytes [2].
   Cold storage of blood leads to aponecrosis of lymphocytes, instead of apoptosis, which happens with fresh cells. Apoptosis leads to an anti-inflammatory effect, suppressing the immune response. However, aponecrosis leads to generation of inflammatory response, leading to detrimental effects on pregnancy, instead of protecting it after LIT [3]. This also explains the higher rate of miscarriages in the treated group [4].
3. Before reaching the initially expected number of participants, the study was ended.
4. Patients with autoimmune diseases (positive ANA tests) were not excluded from the study, which could have adversely affected the outcomes of the immunotherapy using lymphocytes.
5. REMIS allowed LIT to be performed 2 weeks before conception. This was significantly less than the previous study. To allow enough time for the immune response to develop, most studies recommend that LIT be performed at least 4 weeks before planning conception.
6. There was no control of immunomodulation, that was produced after the immunotherapy with lymphocytes, before the couples were allowed to attempt pregnancy.
7. The routes of mononuclear cell administration (intradermal, subcutaneous, intramuscular, or intravenous) were not uniform in all the centers. It is not likely that intramuscular or intravenous routes can generate tolerogenic interactions with dendritic cells. These routes do not provide high concentration of dendritic cells, as found in the skin, for immunological interaction.
8. The number of doses of mononuclear cell injections was not the same in all the cases.
9. The mononuclear cell concentration of the cell preparation used for immunization was not uniform in all the cases.
10. All of the above-suggested factors led to the conclusion that miscarriages among patients receiving immunotherapy with lymphocytes could be caused by an embryonic genetic abnormality, undiagnosed autoimmune disease, or immunotherapy done using insufficient paternal CD200 cells [4].

11.5 Significance of REMIS Study

1. REMIS study was a large, double blind, multicentric, randomized trial.
2. This study was part of the Cochrane review on immunotherapy (including LIT), and it had the highest weightage due to the larger number of patients involved in the study.
3. The REMIS study recommended against the use of LIT in Recurrent miscarriages. Due to this, the Cochrane meta-analysis (2003) showed an Odds Ratio of 1.22 and concluded that LIT has no role in RM.
4. The flaws in the REMIS study were never considered or acknowledged and they continued to impact subsequent Cochrane reviews.
5. With newer meta-analyses, including newer studies, the efficacy of LIT in RM is again highlighted, reducing the impact of the REMIS trial on the recommendations about the role of LIT in RM and RIF.

References

1. Ober C, Karrison T, Odem RB, Barnes RB, Branch DW. Stephenson MD Mononuclear-cell immunization in prevention of recurrent miscarriages: a randomized trial. Lancet. 1999;354:365–9.
2. Kaczmarek A, Vandenabeele P, Krysko DV. Necroptosis: the release of damage-associated molecular patterns and its physiological relevance. Immunity. 2013;38:209–23.
3. Clark DA. Cell-surface CD200 may predict efficacy of paternal mononuclear leukocyte immunotherapy in treatment of human recurrent pregnancy loss. Am J Reprod Immunol. 2009;61:75–84.
4. Clark DA. The end of evidence-based medicine? Inflammopharmacol. 2012;20:187–93.

Cochrane Reviews and LIT

12

Abstract

Cochrane Review is a systematic meta-analysis of research studies. It seeks to recognize, evaluate, and combine all evidence that meets predetermined eligibility requirements for a particular research question. Results from the Cochrane review can be considered reliable evidence which will form the basis of the recommendations and guidelines for this clinical situation. The Cochrane reviews about the role of LIT in RPL/RIF (2003, 2006, 2014) have concluded that LIT has no beneficial effect on prevention of miscarriage and increase in the live births in these clinical conditions. Subsequent studies have pointed out the flaws in these Cochrane reviews. The objections raised by different researchers and their recommendations to correct these deficiencies are discussed in this chapter.

Keywords

Cochrane Database of Systematic Review (CDSR) · Cochrane review · Flaws in Cochrane review

12.1 Introduction

Cochrane Reviews related to LIT have heavily impacted the use of LIT in treatment of RM. To understand more about this, we have to see the Cochrane reviews in detail.

12.1.1 Cochrane Database of Systematic Reviews

Cochrane Database of Systematic Reviews (CDSR) is the most important journal and database that provides systematic reviews in healthcare. CDSR also contains Cochrane Reviews (systematic review) and protocols to conduct Cochrane Reviews,

M. Raut, M. Raut, *Lymphocyte Immunization Therapy (LIT) in Reproductive Failures*, https://doi.org/10.1007/978-981-19-2960-1_12

editorials, and supplements. Cochrane is the producer and owner of CDSR. It is a global self-supporting network that includes professionals, caregivers, patients, and others who have a special interest in healthcare.

12.1.2 Objective of Cochrane Reviews

A Cochrane Review, or systematic review, is put together and supervised by an editorial team of the Cochrane Review Group. It seeks to recognize, evaluate, and combine all evidence that meets predetermined eligibility requirements for a particular research question.

12.2 Cochrane Database of Systematic Reviews for Immunotherapy for Recurrent Miscarriage

A recurring miscarriage refers to three or more miscarriages in a row. Some women may experience a heightened immune response to the embryo, fetus. There have been a variety of therapies that attempt to immunize the woman from the "foreign cells" of future pregnancies. Some immunotherapies include white blood cells (leukocytes), from a woman's partner, or from a donor, products from embryos (trophoblast membranes), and antibodies derived mainly from blood (immunoglobulin).

12.3 Objective

Many immunotherapies can be used to treat a miscarriage. This review examined the effects of various immunotherapies, including intravenous immune globulin and paternal leukocyte vaccination, on live birth rates for women experiencing previous unexplained repeated miscarriages.

12.4 Selection Criteria

Randomized clinical trials of immunotherapies for women with at least three previous miscarriages and only one or no live birth were conducted. Women who had experienced multiple miscarriages had all been excluded.

12.5 Search Strategy

In September 2002, the Cochrane Pregnancy and Childbirth Group Trials Register was searched. The American Society for Reproductive Immunology also provided individual patient data.

12.6 Cochrane Database Reviews About Immunotherapy in Recurrent Miscarriages

1. Cochrane Review 2003 [1]:
   This review was authored by Prof. J. R. Scott, Department of Obstetrics and Gynaecology, University of Utah Medical Centre, Salt Lake City, Utah, USA, and was first published on January 20, 2003.
2. Cochrane Review 2006 [2]:
   This review was authored by Porter TF, LaCoursiere Y, and Scott JR.
3. Cochrane Review 2014 [3]:
   This review was authored by Wong LF, Porter T, and Scott JR.

12.7 Results

1. *Cochrane Review 2003* [1]:
   Nineteen trials of high quality were included.
   Paternal cell immunization (11 trials, 596 women), odds ratio (OR) 1.05, 95% confidence intervals (CI) 0.75–1.47.
   Third-party donor cell immunization (3 trials, 156 women), OR 1.39, 95% CI 0.68–2.82.
   Trophoblast membrane infusion (1 trial, 37 women), OR 0.40, 95% CI 0.11–1.45.
   Intravenous immune globulin (IVIG), OR 0.98, 95% CI 0.61–1.58.
2. *Cochrane 2006* [2]:
   Twenty trials of high quality were included.
   Paternal cell immunization (12 trials, 641 women), Peto odds ratio (Peto OR) 1.23, 95% confidence interval (CI) 0.89–1.70.
   Third-party donor cell immunization (3 trials, 156 women), Peto OR 1.39, 95% CI 0.68–2.82.
   Trophoblast membrane infusion (1 trial, 37 women), Peto OR 0.40, 95% CI 0.11–1.45.
   Intravenous immune globulin, Peto OR 0.98, 95% CI 0.61–1.58.
3. *Cochrane Review 2014*: (see Fig. 12.1) [3]
   The 22 included randomized controlled trials that involved 1137 women and which took place in 11 countries between 1985 and 2004 were included in the review. The four forms of immunotherapy tested, included immunization with white cells from the woman's spouse (12 trials, 641 females), white blood cells from third-party donors (three trials each, 156 women), products made from early embryos (1 trial, 37 women) and antibodies prepared directly from blood (intravenous immune globulin, 8 trials, 303 women).

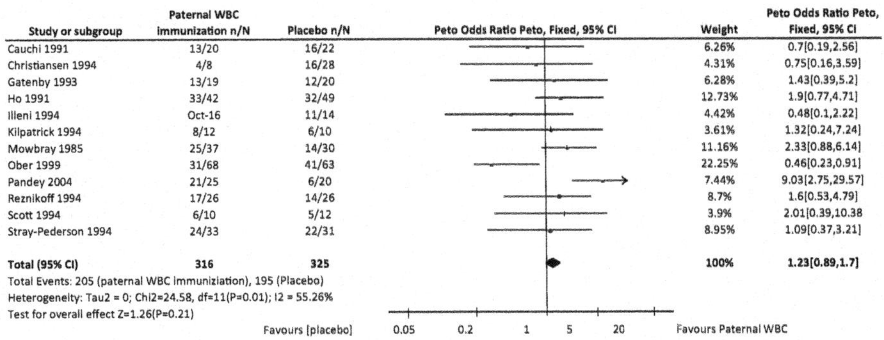

Fig. 12.1 Results of paternal cell immunization [3]

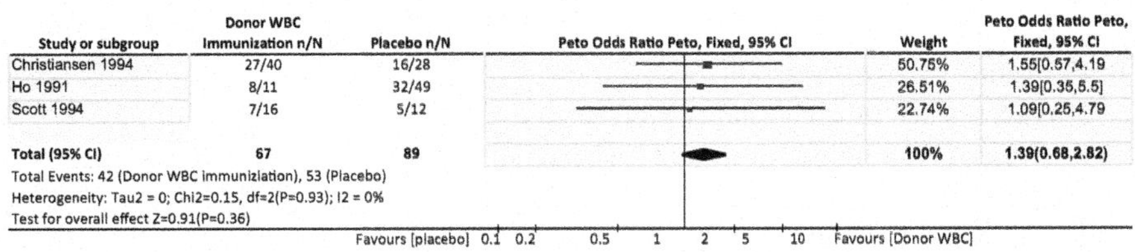

Fig. 12.2 Results of third-party donor immunization [3]

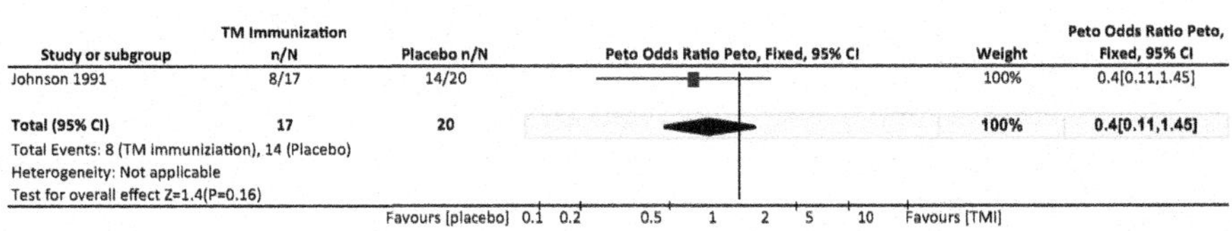

Fig. 12.3 Results of trophoblast membrane infusion [3]

The odds ratio (Peto OR), for paternal cell vaccination (12 trials, 641 females), was 1.23 (95% confidence interval (0.89–1.70), which indicated that LIT is not advantageous as shown in Fig. 12.1 above.

With Third-party donor cell immunization (3 trials, 156 women), the Odds ratio was 1.39, a little better than with paternal lymphocytes (95% CI 0.68–2.82), as shown in Fig. 12.2 [3].

Use of Trophoblast membrane infusion (1 trial, 37 women) showed an Odds ratio of 0.40 (95% CI 0.11–1.45) as shown in Fig. 12.3 [3].

Use of Intravenous immunoglobulin (8 trials, 303 women) showed an Odds Ratio of 0.98 (95% CI 0.61–1.58) as shown in Fig. 12.4 [3].

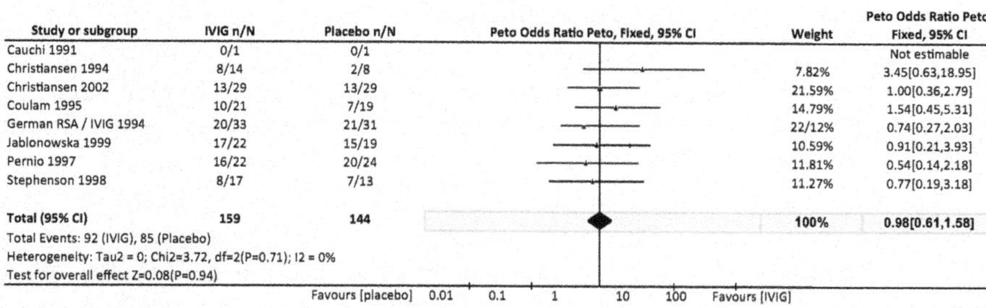

Fig. 12.4 Results of intravenous immunoglobulin infusion [3]

12.8 Effects of Interventions

There were no significant differences in the subsequent live births between immunotherapy treatment groups and the control groups. These include paternal cell vaccination (12 trials, 641 females), Peto odds ratio 1.23, 95% confidence interval 0.89–1.70; third-party donor cell immunization (3 trials, 37 women), Peto OR 1.40, 95%CI 0.11–1.45, and trophoblast membrane injection (one trial, 37 women). Peto OR 0.98; 95%CI 0.61–1.58.

Intravenous immune globulin (IVIG; eight trials, 303 women), Peto OR 0.98; Peto OR 0.98; 95%CI 0.61–1.58, 95%CI 0.61–1.58.

The trials that were published earlier, gave higher estimates of treatment effects, which suggested a publication bias.

Intention-to-treat analysis showed that there were no significant differences in subsequent live births between the treatment and control groups: paternal cell immunization (four trials, 350 women) OR 1.35; 95% CI 0.89–2.05 and IVIG (4 trials, 279 women), OR.18, 95%CI 0.72–1.94.

12.9 Conclusions

Paternal cell immunization, third-party donor leukocytes, trophoblast membranes, and intravenous immune globulin provide no significant beneficial effect over placebo in preventing further miscarriages (2003).

Paternal cell immunization, third-party donor leukocytes, trophoblast membranes, and intravenous immune globulin provide no significant beneficial effect over placebo in improving the live birth rate (2006, 2014).

12.10 Quality of the Evidence

For most of the studies included in the review, there was a low risk of bias.

12.10.1 Criticism of Cochrane Review

The Cochrane Review on the Role of Immunotherapy (including LIT) in Recurrent Miscarriages has stated that it has no beneficial effect compared to the placebo treatment.

The meta-analysis in the Cochrane review itself has been criticized by many researchers, due to various flaws associated with it.

Many subsequent studies have pointed out the deficiencies in the Cochrane Meta-analysis. They have shown that actually in carefully selected cases, immunotherapy can be a valuable tool in the treatment of recurrent spontaneous abortions or implantation failures.

12.10.2 Flaws in Cochrane Reviews (Christiansen O. B.)

A Cochrane meta-analysis of relevant trials found that third-party and paternal LIT does not have a notable advantage over placebo when it comes to preventing miscarriages.

However, if the Cochrane Reviews are studied in detail, it shows the following flaws:

1. The OR for birth post-LIT would be 1.34 (95%CI 0.83–2.15) if all LIT trials were involved in a single meta-analysis. And this is near to significance.
2. A separate analysis of primary RM patients was not performed in the Cochrane analysis. However, LIT was found to be effective in a meta-analysis of primary RM outcomes of 8 RCTs [3].
3. A meta-regression analysis was done in this study and the effect of LIT on previous miscarriages was found to be increased. This suggests that it may be an effect modifier, important in meta-analyses of LIT in RM. This could be explained by the fact that there is an increase in the incidence of immunological risk factor factors with an increase in the number of previous miscarriages.
4. In the Cochrane analysis (2006), there was statistically significant heterogeneity of the outcomes in trials of partner LIT. This raises questions about the logic behind calculating a pooled OR.
5. Despite the fact that the immunizing methods in the included trials were substantially different, the analysis did not distinguish between different immunizing dosages or routes of administration [4].
   (a) Very low and high doses of vaccines usually cause tolerance. However, intermediate doses can induce immunization and cause harm to the fetus. The Cochrane analysis trials showed that one study administered less than 100×10^6 cells and the other more than 1500×10^6 cells.
   (b) Intradermal/subcutaneous injections induce immunization, and on the other hand, IV administration usually induces tolerance. All cells were administered intradermally in one LIT trial. In other trials, however, they were all administered intravenously.

Table 12.1 LIT parameters and Cochrane Review [7]

Patient- or treatment- related parameter	Importance of parameter	Separate analysis in Cochrane review
Primary vs. secondary RM	Yes, in meta-analysis from 1994	No
Few vs. multiple miscarriages	Yes, in meta-analysis from 1994	No
Immunizing dose	Yes, well-known in vaccination	No
Route of administration	Yes, well-known in vaccination	No
Partner vs. donor cells	No evidence that RM is partner-specific	Yes

(c) One trial had cells that were to be injected, stored overnight, while in other trials, the cells were injected within a few hours of being obtained from the donor/partner. Lymphocyte storage can alter the immunogenetic properties of lymphocytes, which could lead to a decrease in anti-abortive effects.

(d) LIT with donor cells or partner cells can increase the live birth rate by 34% according to a meta-analysis. So instead of rejecting immunotherapy's efficacy, the Cochrane review should have concluded that certain types of immunotherapy in RM are promising, but more RCTs are required. There should be new meta-analyses done according to the dose of the immunization agent and route of administration. The number of miscarriages and the type of PRL (primary or secondary) should also be taken into account [5–7].

Table 12.1 shows the importance and differences in LIT parameters. It also indicates whether separate analyses of these parameters were performed in the Cochrane reviews (Dr. O. B. Christiansen) [7] (see Table 12.1).

New Meta-analyses of studies are related to the Role of Immunotherapy in Recurrent miscarriages have been discussed in the next chapter.

References

1. Scott JR. Cochrane Database of Systematic Reviews Immunotherapy for recurrent miscarriage; January 2003. https://doi.org/10.1002/14651858.CD000112
2. Porter TF, La Coursiere Y, Scott JR. Immunotherapy for recurrent miscarriage [Cochrane Review]. Database Systematic Rev. 2006(2):CD000112. https://doi.org/10.1002/14651858.CD000112
3. Wong LF, Flint Porter T, Scott JR. Immunotherapy for recurrent miscarriage. Cochrane Database Syst Rev. 2014;2014(10):CD000112. https://doi.org/10.1002/14651858.CD000112.pub3.

4. Daya S, Gunby J, RMITG. The effectiveness of allogeneic leukocyte immunization in unexplained primary recurrent spontaneous abortion. Am J Reprod Immunol. 1994;32:294–302.
5. Christiansen OB, Nielsen HS. Intravenous immunoglobulin in the prevention of recurrent miscarriage: does it work? In: Markert UR, editor. Immunology of gametes and embryo implantation. Chem Immunol Allergy, vol. 88. Basel: Karger; 2005. p. 117–27.
6. Kling C, Steinmann J, Westphal E, Magez J, Kabelitz D. Adverse effects of intradermal allogeneic lymphocyte immunotherapy: acute reactions and role of autoimmunity. Hum Reprod. 2006;21:429–35.
7. www.comtecmed.com/COGI/COGI9/Uploads/assets/christiansen%20ob.pdf

Recent Meta-Analyses and LIT

13

Abstract

Cochrane Reviews on LIT's role in RPL/RIF have shown no benefits. However, subsequent studies pointed out the deficiencies in the Cochrane reviews. The study with maximum weightage in the Cochrane reviews (REMIS study) had methodological flaws, indicating that the results of the REMIS trial were incorrect. A new meta-analysis of studies incorporated in the Cochrane review omitting REMIS study showed better results in favor of LIT. Newer meta-analyses (Liu et al. and Cavalcante et al.), that included newer studies along with the studies in the Cochrane reviews, showed a significantly high odds ratio, proving the efficacy of LIT in RPL/RIF. It was also shown that LIT is more beneficial when it is given, both before and during pregnancy.

Keywords

Cochrane review · Ober study · Liu meta-analysis · Cavalcante meta-analysis · LIT in pregnancy

13.1 Introduction

The Cochrane Review on Role of Immunotherapy (including LIT) in Recurrent Miscarriages has stated that LIT has no beneficial effect compared to the placebo treatment. Many subsequent studies have pointed out the deficiencies in the Cochrane Meta-analysis and have shown that actually Immunotherapy can play a beneficial role in recurrent spontaneous abortions and implantation failures, in properly selected cases.

M. Raut, M. Raut, *Lymphocyte Immunization Therapy (LIT) in Reproductive Failures*, https://doi.org/10.1007/978-981-19-2960-1_13

13.2 Need for Newer Meta-Analyses

In general, LIT was found to be highly successful in couples who have had multiple miscarriages in most studies from the 1980s and 1990s. However, many of these studies were not randomized controlled trials. Very few of these studies were conducted using randomized controlled trials (RCT).

This has been described in detail in the chapter on the history of LIT. The first high-quality RCT was performed by Mowbray (1985) and it showed a success rate of 77.3% compared to 37% in the placebo group.

The Ober study, also known as the REMIS study, that was published in 1999 showed that LIT was not useful in Recurrent Miscarriages (see Chap. 10). Later, many studies were published related to the efficacy of LIT in RM and RIF. In 2010, a meta-analysis of randomized trials of LIT was published by Flint PT et al. The meta-analysis showed that both immunizations with maternal leukocytes and immunizations using third-party donors had no effect on the live birth rate of URSA affected women. It concluded that, LIT should not be suggested as a treatment of URSA. The Cochrane reviews about the role of LIT in URSA were based mainly on Ober study. Subsequent articles and papers had shown the flaws in the Ober study and had proved how the results of the Ober study were unreliable. All the subsequent recommendations and guidelines related to use of LIT in URSA were based on the recommendations of Ober study. Also, by that time, there were many studies published proving that LIT is effective in preventing miscarriage in URSA [1].

13.3 New Meta-Analysis of Cochrane Review Data

Cochrane Database of Systematic Reviews conducted a meta-analysis of 12 studies related to the role of LIT in URSA. Of these 12 studies, the maximum weightage was to the study published in Lancet (1999) by Ober et al. However, the Ober study, as it is known, has faced a lot of criticism. Many subsequent reviews and reports about the Ober study highlighted the methodological flaws in the study, indicating that the results of the Ober trial were incorrect (see Chap. 10). Therefore, a new meta-analysis was done of all the studies included in the Cochrane Review, omitting the Ober study.

This new meta-analysis found a marked improvement in live births in couples who underwent LIT. The OR was 1.63, as opposed to 1.23 in Cochrane's original analysis. When the results of Ober et al. were omitted, it resulted in greater homogeneity than the original Cochrane analysis (see Fig. 13.1) [2].

Study or Subgroup	LIT Event	Total	Placebo Events	Total	Weight	Odds Ratio M.H. Fixed 95% CL	Year
Mowbray 1985	25	37	14	30	11.30%	2.38 [0.88, 6.43]	1985
Cauchi 1991	13	20	16	22	12.00%	0.70 [0.19, 2.59]	1991
Ho 1991	33	42	32	49	14.20%	1.95 [0.76, 5.00]	1991
Gatenby 1993	13	19	12	20	8.30%	1.44 [0.39, 5.39]	1993
Christiansen 1994	4	8	16	28	8.00%	0.75 [0.16, 3.62]	1994
Scott 1994	6	10	5	12	4.1%	2.10 [0.38, 11.59]	1994
Kilpatrick 1994	8	12	6	10	4.9%	1.33 [0.23, 7.63]	1994
Reznikoff 1994	17	26	14	26	10.9%	1.62 [0.53, 4.95]	1994
Stray -Pederson 1994	24	33	22	31	13.9%	1.09 [0.37, 3.25]	1994
Illeni 1994	10	16	11	14	9.9%	0.45 [0.09, 2.32]	1994
Ober 1999	31	68	41	63	0.0%	0.45 [0.22, 0.91]	1999
Pandey 2004	21	25	6	20	2.4%	12.25 [2.92, 51.42]	2004
Total (95% CI)		**248**		**262**	**100%**	**1.63 [1.13, 2.35]**	
Total events	174		154				

Heterogeneity. Chi2 = 13.88, df = 10 (P = 0.18), I2 = 28%
Test for overall effect: Z = 2.60 (P= 0.009)

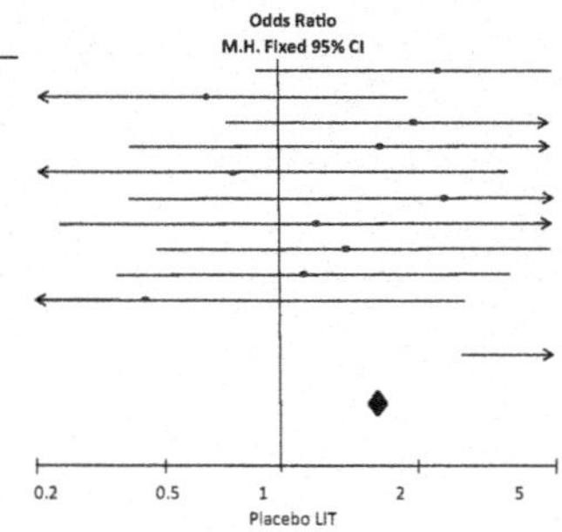

Fig. 13.1 Statistics from Cochrane Meta-analysis on LIT and RM. Data by Ober et al. has been removed [2]

13.4 New Meta-Analysis of Studies Related to Role of LIT in uRPL (Liu Z et al. 2016) [2]

The Cochrane meta-analysis of studies related to the role of LIT in uRPL had many flaws (see Chap. 11). Liu et. al. published a revised meta-analysis in the *American Journal of Reproductive Immunology*. The meta-analysis's purpose was to provide the latest evidence on the subject and rectify the flaws in Cochrane's meta-analysis. The meta-analysis included 18 randomized placebo-controlled clinical research studies. These were all conducted between 1985 and 2013. The trial had 1738 participants, 739 in the immunization program with paternal or unrelated donors, and 999 in the control.

The important features of the meta-analysis by Liu et al.:

1. Liu et al. did not include data from Ober et al. This was due to the methodological flaws in the literature. The data by Stray-Pederson et al. was also not included, because it did not appear in their search base.
2. Liu et al. demonstrated that lymphocyte immunization can increase live birth rates to 77.8% compared with 46.1% for the control group, with an OR of 4.02 (see Fig. 13.2) [2], and CIs of 3.23–5.00 (Fig. 13.2). The meta-analysis by Liu et al. shows that the data presented by Christiansen et al. are consistent with what was there in the original publication.
   When the data of LIT and placebo given to women with URSA was compared, there is OR of 3.74 and CI of 3.07–4.57 (see Fig. 13.3) [2].
3. If we analyze the data of Liu et al. separately, depending on whether the donor of lymphocytes is paternal or unrelated, the beneficial effects of immunotherapy with lymphocytes in preventing miscarriages, are still evident.
4. Liu et al. conducted statistical analyses of subgroups of patients according to clinical protocols, which helped to determine which protocol for immunization with lymphocytes yields the best results.

Study or Subgroup	LIT Event	Total	Placebo Events	Total	Weight	Odds Ratio M.H. Fixed 95% CL	Year
Mowbray 1985	25	37	14	30	5.9%	2.38, [0.88,6.43]	1985
Ho 1991	33	42	32	49	7.5%	1.95 [0.76,5.00]	1991
Cauchi 1991	13	20	16	22	6.3%	0.70 [0.19, 2.59]	1991
Clark 1991	7	11	2	7	1.1%	4.38 [0.56, 33.95]	1991
Gatenby 1993	13	19	12	20	4.4%	1.44 [0.39, 5.39]	1993
Scott 1994	6	10	5	12	2.2%	2.10 [0.38, 11.59]	1994
Reznikoff 1994	17	26	14	26	5.7%	1.62 [0.53, 4.95]	1994
Ileni 1994	10	16	11	14	5.2%	0.45 [0.09, 2.32]	1994
Kilpatrick 1994	8	12	6	10	2.6%	1.33 [0.23, 7.63]	1994
Christiansen 1994	31	48	16	28	8.5%	1.37 [0.53, 3.55]	1994
Carp 1997	5	11	11	31	3.7%	1.52 [0.37, 6.12]	1997
Pandey 2002	12	14	1	5	0.2%	24.00 [1.69, 340.99]	2002
Hong 2003	18	21	2	8	0.5%	18.00 [2.40, 134.83]	2003
Pandey 2004	21	25	6	20	1.3%	12.25 [2.92, 51.42]	2004
Yanping 2011	41	49	24	45	4.8%	4.48 [1.72, 11.68]	2011
Lin 2012	33	42	17	42	4.3%	5.39 [2.06, 14.09]	2012
Aiwu 2013	32	39	18	39	3.8%	5.33 [1.90, 14.97]	2013
Bin 2013	250	297	254	591	31.9%	7.06 [4.97, 10.03]	2013
Total (95% CI)		**739**		**999**	**100%**	**4.02 [3.23, 5.00]**	
Total events	575		461				

Heterogeneity. Chi2 = 47.53, df = 17 (P = 0.00001), I2 = 64%
Test for overall effect: Z = 12.43 (P <0.001)

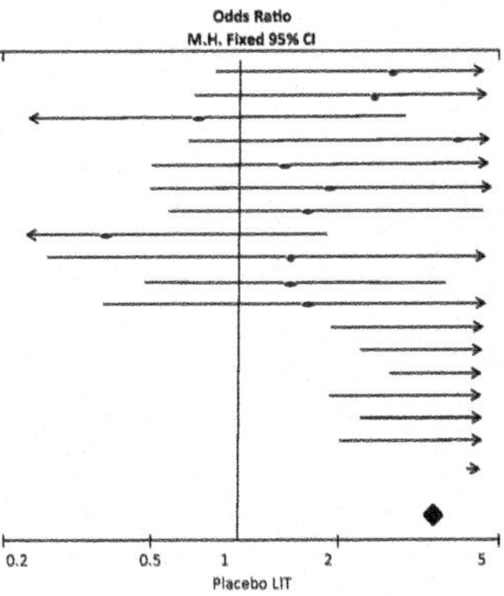

Fig. 13.2 Statistics of the meta-analysis by Liu et al. [2]

Study or Subgroup	LIT Event	Total	Placebo Events	Total	Weight	Odds Ratio M.H. Fixed 95% CL	Year
Mowbray JF 1985	25	37	14	30	4.2%	2.33 [0.88, 6.14]	1985
HoHN 1991	33	42	32	49	4.8%	1.90 [0.77, 4.71]	1991
Clark DA 1991	7	11	2	7	1.2%	3.76 [0.50, 23.73]	1991
Cauchi MN 1991	13	20	16	22	2.4%	0.70 [0.39, 5.20]	1991
Gatenby PA 1993	13	19	12	20	2.4%	1.43 [0.39, 5.20]	1993
Ileni MT 1994	10	16	11	14	1.7%	0.48 [0.10, 2.22]	1994
Kilpatrick DC 1994	8	12	6	10	1.4%	1.32 [0.24, 7.24]	1994
Reznikoff ET MF 1994	17	26	14	26	3.3%	1.60 [0.53, 4.79]	1994
Scott JR 1994	6	10	5	12	1.5%	2.01 [0.39, 10.38]	1994
Christiansen CB 1994	31	48	16	28	4.4%	1.37 [0.53, 3.54]	1994
Carp HJ 1997	5	11	11	31	2.0%	1.51 [0.37,6.12]	1997
Pandey MK 2002	12	14	1	5	0.9%	17.84 [2.10, 151.33]	2002
Hong L 1991	18	21	2	8	1.3%	15.47 [2.74, 87.22]	2003
Pandey MK 2004	21	25	6	20	2.8%	9.03 [2.75, 29.57]	2004
Yanping C 2011	41	49	24	45	5.2%	4.08 [1.71, 9.76]	2011
Lin S 2012	33	42	17	42	5.3%	4.77 [2.01, 11.34]	2012
Bin T 2013	32	39	18	39	4.7%	4.66 [1.86, 11.70]	2013
Aiwu W 2013	250	297	244	591	50.4%	5.35 [4.04, 7.08]	2013
Total (95% CI)		**739**		**999**	**100%**	**3.74 [3.07, 4.57]**	
Total events	**575**		**461**				

Heterogeneity. Chi2 = 42.19, df = 17 (P = 0.00006), I2 = 60%
Test for overall effect: Z = 12.96 (P < 0.00001)

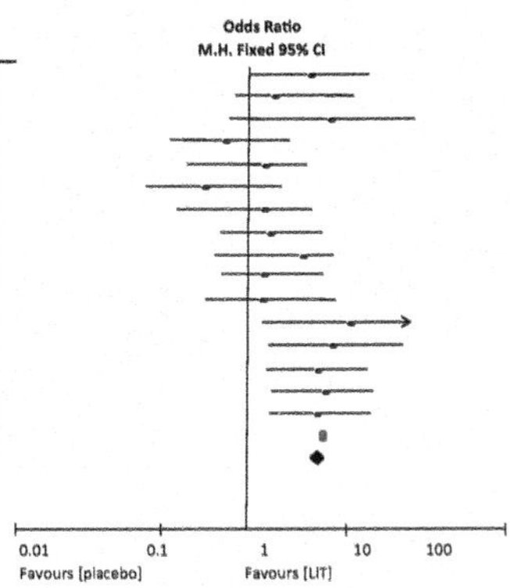

Fig. 13.3 Comparison of LIT and placebo for URSA—Liu et al. [2]

(a) Timing of treatment:
It was determined whether immunotherapy was administered only before or during pregnancy. Both types of treatment substantially increased the number of live births. The results were much better when the immunization was done before and during pregnancy (see Figs. 13.4 and 13.5) [2].

(b) Concentration of lymphocytes
When the LIT contained fewer than 100×10^6 lymphocytes per application, the results were more favorable than when the lymphocyte concentration was higher than 100×10^6 per application (see Figs. 13.6 and 13.7) [2].

(c) Effect of LIT on Asian patients
It was seen that the success rate of LIT in Asian patients was significantly higher when compared to European and American patients (See Figs. 13.8 and 13.9) [2].

Study or Subgroup	LIT Event	Total	Placebo Events	Total	Weight	Odds Ratio M.H. Fixed 95% CL	Year
Mowbray JF 1985	25	37	14	30	14.2%	2.33 [0.88,6.14]	1985
Cauchi MN 1991	13	20	16	22	8.0%	0.70 [0.19, 2.65]	1991
Clark DA 1991	7	11	2	7	3.9%	3.76 [0.60, 23.73]	1991
Ho HN 1991	33	42	32	49	16.2%	1.90 [0.77, 4.71]	1991
Gatenby PA 1993	13	19	12	20	8.0%	1.43 [0.39, 5.20]	1993
Christiansen CB 1994	27	40	16	28	13.5%	1.55 [0.57, 4.19]	1994
Scott JR 1994	6	10	5	12	5.0%	2.01 [0.39, 10.38]	1994
Reznikoff ET MF 1994	7	10	14	26	6.3%	1.91 [0.44, 8.21]	1994
Illeni MT 1994	10	16	11	14	5.6%	0.48 [0.10, 2.22]	1994
Carp HJ 1997	5	11	11	31	6.8%	1.51 [0.37, 6.12]	1997
Pandey MK 2002	12	14	1	5	2.9%	17.84 [2.10, 151.33]	2002
Pandey MK 2004	21	25	6	20	9.5%	9.03 [2.75, 29.57]	2004
Total (95% CI)		**255**		**264**	**100%**	**2.00[1.39, 2.88]**	
Total events	**179**		**140**				

Heterogeneity. Chi2 = 17.29, df = 11 (P = 0.10), I2 = 36%
Test for overall effect: Z = 3.71 (P = 0.0002)

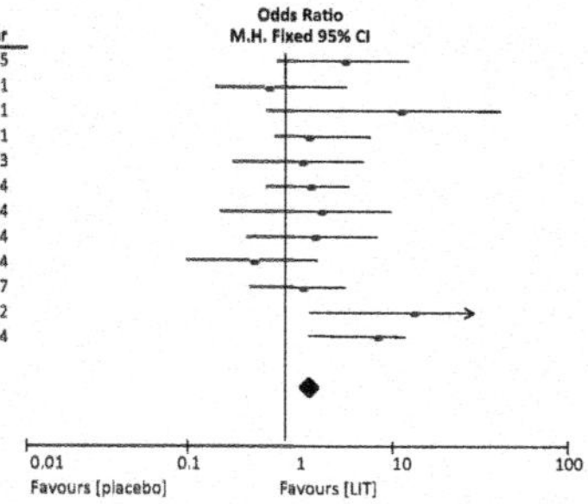

Fig. 13.4 Comparison of LIT with placebo for URSA prior to pregnancy only [2]

Study or Subgroup	LIT Event	Total	Placebo Events	Total	Weight	Odds Ratio M.H. Fixed 95% CL	Year
Christiansen CB 1994	4	8	16	28	2.3%	0.75 [0.16,3.59]	1934
Kipatrick DC 1994	8	12	6	10	1.8%	1.32 [0.24, 7.24]	1994
Reznikoff ET MF 1994	10	16	14	26	3.6%	1.41 [0.41, 4.90]	1994
Hong L 1991	18	21	2	8	1.8%	15.47 [2.74, 87.22]	2003
Yanping C 2011	41	49	24	45	7.2%	4.08 [1.71, 9.76]	2011
Lin S 2012	33	42	17	42	7.3%	4.77 [2.01, 11.34]	2012
Bin T 2013	32	39	18	39	6.5%	4.66 [1.86, 11.70]	2013
Aiwu W 2013	250	297	254	591	69.4%	5.35 [4.04, 7.08]	2013
Total (95% CI)		**484**		**789**	**100%**	**4.67 [3.70, 5.90]**	
Total events	**396**		**351**				

Heterogeneity. Chi2 = 13.75, df = 7 (P = 0.06), I2 = 49%
Test for overall effect: Z = 12.89 (P < 0.00001)

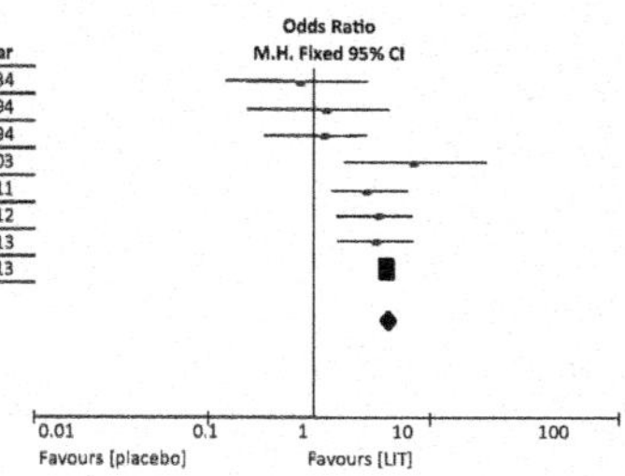

Fig. 13.5 Comparing of LIT versus placebo for URSA during and before pregnancy [2]

Study or Subgroup	LIT Event	Total	Placebo Events	Total	Weight	Odds Ratio M.H. Fixed 95% CL	Year
Mowbray JF 1985	25	37	14	30	15.6%	2.33 [0.88, 6.14]	1985
Cauchi MN 1991	13	20	16	22	8.7%	0.70 [0.19, 2.56]	1991
Clark DA 1991	7	11	2	7	4.3%	3.76 [0.60, 23.73]	1991
Ho HN 1991	33	42	32	49	17.6%	1.90 [0.77, 4.71]	1991
Gatenby PA 1993	13	19	12	20	8.8%	1.43 [0.39, 5.20]	1993
Christiansen CB 1994	31	48	16	28	16.1%	1.37 [0.53, 3.54]	1994
Scott JR 1994	6	10	5	12	5.4%	2.01 [0.39, 10.38]	1994
Reznikoff ET MF 1994	17	26	14	26	12.1%	1.60 [0.53, 4.79]	1994
Illeni MT 1994	10	16	11	14	6.2%	0.48 [0.10, 2.22]	1994
Kilpatrick DC 1994	8	12	6	10	5%	1.32 [0.24, 7.24]	1994
Total (95% CI)		**241**		**218**	**100%**	**1.52 [0.04, 2.22]**	
Total events	**163**		**126**				

Heterogeneity. Chi2 = 5.66, df = 9 (P = 0.77), I2 = 0%
Test for overall effect: Z = 2.14 (P = 0.03)

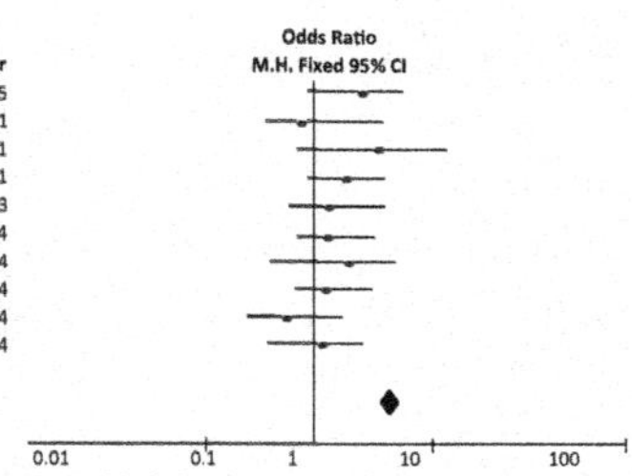

Fig. 13.6 Comparing high doses LIT versus placebo for URSA [2]

Study or Subgroup	LIT Event	Total	Placebo Events	Total	Weight	Odds Ratio M.H. Fixed 95% CL	Year
Carp HJ 1997	5	11	11	31	2.8%	1.51 [0.37, 6.12]	1997
Pandey MK 2002	12	14	1	5	1.2%	17.84 [2.10, 151.33]	2002
Hong L 1991	18	21	2	8	1.8%	15.47 [2.74, 87.22]	2003
Pandey MK 2004	21	25	6	20	3.9%	9.03 [2.75, 29.57]	2004
Yanping C 2011	41	49	24	45	7.2%	4.08 [1.7, 19.76]	2011
Lin S 2012	33	42	17	42	7.3%	4.77 [2.01, 11.34]	2012
Aiwu W 2013	32	39	18	39	6.5%	4.66 [1.86, 11.70]	2013
Bin T 2013	250	297	254	591	69.3%	5.35 [4.04, 7.08]	2013
Total (95% CI)		**498**		**781**	**100%**	**5.25 [4.16, 6.64]**	
Total events	**412**		**333**				

Heterogeneity. Chi2 = 7.05, df = 9 (P = 0.42), I2 = 1%
Test for overall effect: Z = 13.89 (P <0.00001)

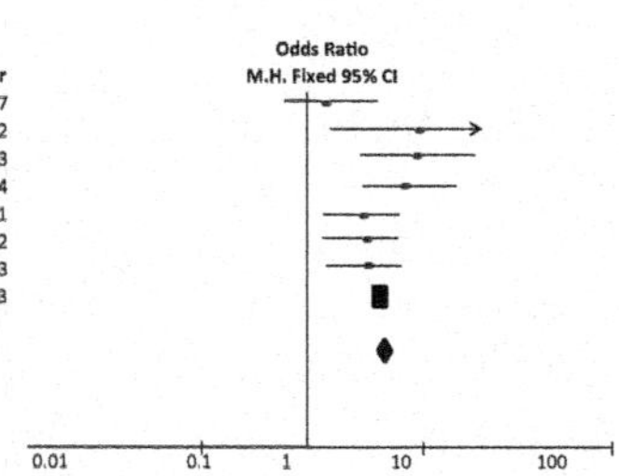

Fig. 13.7 Comparing low doses of LIT and placebo for URSA [2]

Study or Subgroup	LIT Event	LIT Total	Placebo Events	Placebo Total	Weight	Odds Ratio M.H. Fixed 95% CL	Year
HoHN 1991	33	42	32	49	6.4%	1.90 [0.77, 4.71]	1991
Pandey MK 2002	12	14	1	5	1.2%	17.84 [2.10, 151.33]	2002
Hong L 2003	18	21	2	8	1.8%	15.47 [2.74, 87.22]	2003
Pandey MK 2004	21	25	6	20	3.7%	9.03 [2.75, 29.57]	2004
Yanping C 2011	41	49	24	45	6.9%	4.08 [1.71, 9.76]	2011
Lin S 2012	33	42	17	42	7.0%	4.77 [2.01, 11.34]	2012
Bin T 2013	32	39	18	39	6.2%	4.66 [1.86, 11.70]	2013
Aiwu W 2013	250	297	254	591	66.7%	5.35 [4.04, 7.08]	2013
Total (95% CI)		**529**		**799**	**100%**	**5.09 [4.05, 6.40]**	
Total events	**440**		**354**				

Heterogeneity. Chi2 = 8.75, df = 7 (P = 0.27), I2 = 20%
Test for overall effect: Z = 13.89 (P < 0.00001)

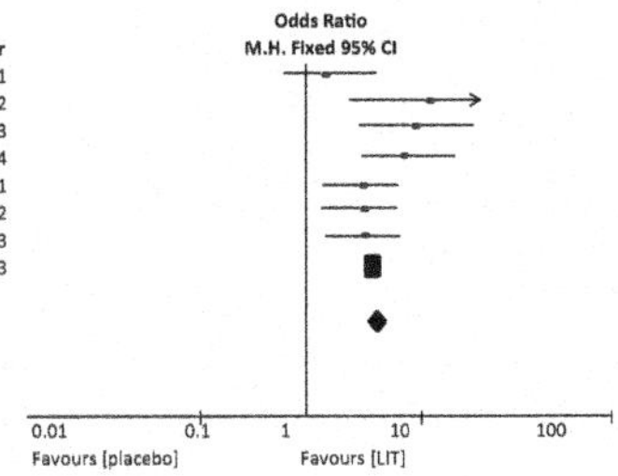

Fig. 13.8 Success of LIT for Asian patients [2]

Study or Subgroup	LIT Event	LIT Total	Placebo Events	Placebo Total	Weight	Odds Ratio M.H. Fixed 95% CL	Year
Mowbray JF 1985	25	37	14	30	17.4%	2.33 [0.88, 6.14]	1985
Clark DA 1991	7	11	2	7	4.8%	3.76 [0.60, 23.73]	1991
Cauchi MN 1991	13	20	16	22	9.7%	0.70 [0.19, 2.56]	1991
Gatenby PA 1993	13	19	12	20	9.8%	1.43 [0.39, 5.20]	1993
Reznikoff ET MF 1994	17	26	14	26	13.5%	1.60 [0.53, 4.79]	1994
Scott JR 1994	6	10	5	12	6.1%	2.01 [0.39, 10.38]	1994
Kilpatrick DC 1994	8	12	6	10	5.6%	1.32 [0.24, 7.24]	1994
Illeni MT 1994	10	16	11	14	6.9%	0.48 [0.10, 2.22]	1994
Christiansen CB 1994	31	48	16	28	17.9%	1.37 [0.53, 3.54]	1994
Carp HJ 1997	5	11	11	31	8.3%	1.51 [0.37, 6.12]	1997
Total (95% CI)		**210**		**200**	**100%**	**1.45 [0.97, 2.17]**	
Total events	**135**		**107**				

Heterogeneity. Chi2 = 5.38, df = 9 (P = 0.80), I2 = 0%
Test for overall effect: Z = 1.80 (P = 0.07)

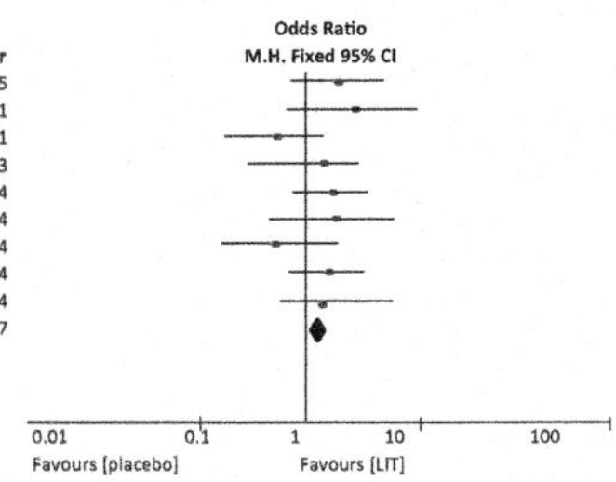

Fig. 13.9 Effect of LIT for European and American patients [2]

13.4.1 Conclusion

The positive effects of LIT were confirmed by the meta-analysis and it concluded that LIT should be performed before and during pregnancy. It also highlighted the need to study the safety, feasibility, and mechanism of LIT as well as to standardize inclusion/exclusion criteria [2].

13.5 New Meta-Analysis of Studies Related to Role of LIT in Recurrent Miscarriages (Cavalcante MB et al., December 2016) [3]

13.5.1 Introduction

The meta-analysis of studies related to the role of LIT in URSA, published by Liu et al. in March 2016, proved the efficacy of LIT in URSA. The subgroup analysis done in the meta-analysis by Liu also showed that LIT is effective in certain subgroups (see above). However, the meta-analysis by Liu recommended investigating the safety and feasibility of LIT. Cavalcante MB et al. performed a systematic review as well as a meta-analysis in December 2016 which contained 6 meta-analyses and 20 randomized clinical trials.

13.5.2 Systematic Review of Meta-Analyses

Two of the six published meta-analyses showed no improvement in live birth rates when LIT was used with RM. The remaining four meta-analyses showed a positive effect of LIT, with a significant increase in the rate of live birth (see Table 13.1).

13.5.3 Meta-Analysis of Randomized Clinical Trials

A meta-analysis of 20 publications of randomized clinical trials of LIT in RM was done. It even included Ober study, which was excluded by Liu in his meta-analysis. Results showed that LIT was associated with an odds ratio of 3.13, indicating that LIT led to a significant increase in live births for couples with RM (see Fig. 13.10) [3].

Data from Liu et al. was also studied separately, to determine the efficacy of paternal lymphocytes compared to allogenic donor lymphocytes. The results showed that miscarriages were significantly reduced by using only paternal Lymphocytes also (see Fig. 13.11) [3].

Table 13.1 List of meta-analysis in the Systematic Review (Cavalcante MB et al.)

Study	Journal	Observations
Wong LF et al., 2013	Cochrane Database Syst Rev(10): CD000112	No improvement in live birth rate
Fraser EJ et al., 1993	Obstet Gynecol. 82:854–859	No improvement in live birth rate
Collins J et al., 1994	Am J Reprod Immunol 32:275–280	Significant improvement in live birth rate
Coulam CB et al., 1994	Am J Reprod Immunol 32:55–72	Significant improvement in live birth rate
Pandey MK et al., 2004	Arch Gynecol Obstet 269:161–172	Significant improvement in live birth rate
Liu Z et al., 2016	Am J Reprod Immunol 76:443–453	Significant improvement in live birth rate

Study or Subgroup	LIT Event	Total	Placebo Events	Total	Weight	Odds Ratio M.H. Fixed 95% CL	Year
Mowbray JF 1985	25	37	14	30	4.4%	2.38 [0.88, 6.43]	1985
Cauchi MN 1991	13	20	16	22	4.7%	0.70 [0.19, 2.59]	1991
Ho 1991	33	42	32	49	5.6%	1.95 [0.76, 5.00]	1991
Clark 1991	7	11	2	4	0.8%	4.38 [0.56, 33.95]	1991
Gatenby 1993	13	19	12	20	3.2%	1.44 [0.39, 5.39]	1993
Christiansen 1994	31	48	16	28	6.3%	1.37 [0.53, 3.55]	1994
Reznikoff 1994	17	26	14	26	4.3%	1.62 [0.53, 4.95]	1994
Scott 1994	6	10	5	12	1.6%	2.10 [0.38, 11.59]	1994
Stray-Pederson 1994	24	33	22	31	5.4%	1.09 [0.37, 3.25]	1994
Kilpatrick 1994	8	12	6	10	1.9%	1.33 [0.23, 7.63]	1994
Illeni 1994	10	16	11	14	3.9%	0.45 [0.09, 2.32]	1994
Carp 1997	5	11	11	31	2.8%	1.52 [0.37, 6.12]	1997
Ober 1999	31	68	41	63	20.4%	0.45 [0.22, 0.91]	1999
Pandey 2002	12	14	1	5	0.2%	24.00 [1.69, 340.99]	2002
Hong 2003	18	21	2	8	0.4%	18.00 [2.40, 134.83]	2003
Pandey 2004	21	25	6	20	0.9%	12.25 [2.92, 51.42]	2004
Yanping 2011	41	49	24	45	3.6%	4.48 [1.72, 11.68]	2011
Lin 2012	33	42	17	42	3.2%	5.39 [2.06, 14.09]	2012
Alwu 2013	250	297	254	591	23.6%	7.06 [4.97, 10.03]	2013
Bin 2013	32	39	18	39	2.8%	5.33 [1.90, 14.97]	2013
Total (95% CI)		**840**		**1093**	**100%**	**3.13 [2.56, 3.82]**	
Total events	**630**		**524**				

Heterogeneity. Chi2 = 84.23, df = 19 (P = 0.00001), I2 = 77%
Test for overall effect: Z = 11.20 (P < 0.00001)

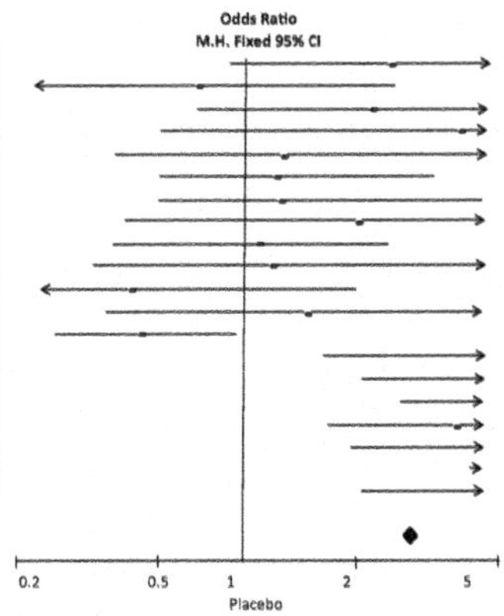

Fig. 13.10 Statistics that includes the results of Cochrane, Liu et al. Meta-analyses on the effect of LIT in RM [3]

Study or Subgroup	LIT Event	Total	Placebo Events	Total	Weight	Odds Ratio M.H. Fixed 95% CL	Year
Mowbray JF 1985	25	37	14	30	13.2%	2.38 [0.88, 6.43]	1985
Ho 1991	33	42	32	49	0.0%	1.95 [0.76, 5.00]	1991
Cauchi MN 1991	13	20	16	22	14.1%	0.70 [0.19, 2.59]	1991
Clark 1991	7	11	2	7	2.3%	4.38 [0.56, 33.95]	1991
Gatenby 1993	13	19	12	20	9.8%	1.44 [0.39, 5.39]	1993
Scott 1994	6	10	5	12	0.0%	2.10 [0.38, 11.59]	1994
Reznikoff 1994	17	26	14	26	12.8%	1.62 [0.53, 4.95]	1994
Illeni 1994	10	16	11	14	11.6%	0.45 [0.09, 2.32]	1994
Kilpatrick 1994	8	12	6	10	5.8%	1.33 [0.23, 7.63]	1994
Christiansen 1994	31	48	16	28	0.0%	1.37 [0.53, 3.55]	1994
Carp 1997	5	11	11	31	8.3%	1.52 [0.37, 6.12]	1997
Pandey 2004	12	14	1	5	0.6%	24.00 [1.69, 340.99]	2002
Hong 2003	18	21	2	8	1.1%	18.00 [2.40, 134.83]	2003
Pandey 2004	21	25	6	20	0.0%	12.25 [2.92, 51.42]	2004
Yanping 2011	41	49	24	45	10.8%	4.48 [1.72, 11.68]	2011
Lin 2012	33	42	17	42	9.6%	5.39 [2.06, 14.09]	2012
Aiwu 2013	32	39	18	39	0.0%	5.33 [1.90, 14.97]	2013
Bin 2013	250	297	254	591	0.0%	7.06 [4.97, 10.03]	2013
Total (95% CI)		**278**		**260**	**100%**	**2.45 [1.71, 3.52]**	
Total events	**202**		**130**				

Heterogeneity. Chi2 = 20.74, df = 11 (P = 0.04), I2 = 47%
Test for overall effect: Z = 4.88 (P < 0.00001)

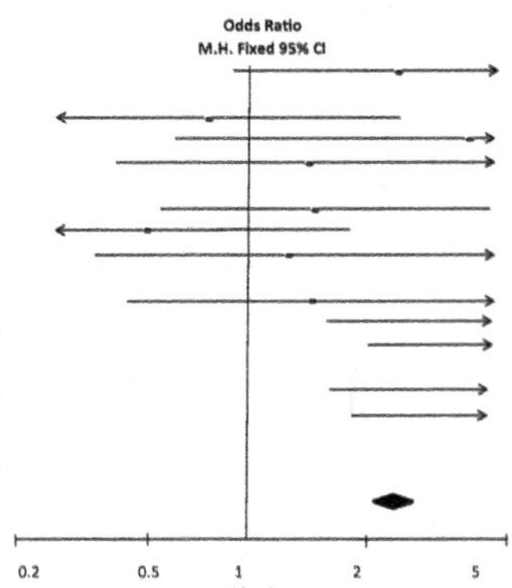

Fig. 13.11 Statistics by Liu et al. Meta-analysis of the effects of LIT on RM, as quoted by Cavalcante et al. [3]

13.5.4 Conclusion

1. The study concluded that LIT is a safe treatment.
2. It further stated that LIT is a valid treatment for couples with RM, when performed with lymphocytes from fresh, non-stored blood.
3. It is beneficial when given before and during pregnancy.
4. The best results are seen when LIT is done with adequate number of lymphocytes.
5. LIT is the most effective when applied intradermally.
6. LIT is beneficial in well-selected couples (subgroups).

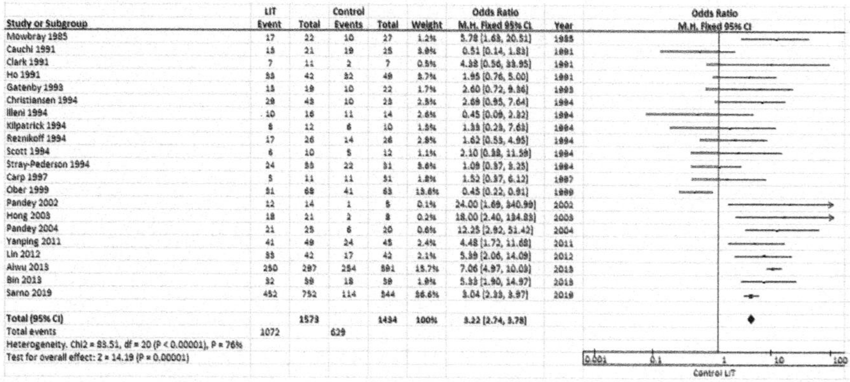

Fig. 13.12 Recent meta-analysis (Cavalcante et al. [4])

13.6 Recent Meta-Analysis by Cavalcante et al. [4]

The systematic review and meta-analysis was recently updated by Cavalcante et al. [4]. This meta-analysis included all the studies reported in the two most relevant meta-analyses together with the data from a large multicentric clinical trial published by Cavalcante et al. This study had taken into consideration other co-existing causes of recurrent miscarriages (autoimmunity and thrombophilia), and they were corrected appropriately before randomization. This meta-analysis had 3007 participants with 1573 in the LIT arm and 1434 in the control arm. The outcomes revealed that women with LIT had an increased rate of live birth than the control arm with OR 3.22 (95% CI 2.74–3.78, $p < 0.00001$) (Fig. 13.12) [4].

References

1. Wong LF, Porter TF, Scott JR. Immunotherapy for recurrent miscarriage. Cochrane Database Syst Rev. 2014;(10):CD000112.
2. Liu Z, Xu H, Kang X, Wang T, He L, Zhao A. Allogenic lymphocyte immunotherapy for unexplained recurrent spontaneous abortion: a meta-analysis. Am J Reprod Immunol. 2016;76: 443–53.
3. Cavalcante MB, Sarno M, Edward AJ, Costa FDS, Barini R. Lymphocyte immunotherapy in the treatment of recurrent miscarriage: systematic review and meta-analysis. Arch Gynaecol Obstet. 2016;295(2):511–8. https://doi.org/10.1007/s00404-016-4270-z.
4. Cavalcante MB, Sarno M, Barini R. Lymphocyte immunotherapy in recurrent miscarriage and recurrent implantation failure. Am J Reprod Immunol. 2021;85(4):e13408. https://doi.org/10.1111/aji.13408.

Review of International Studies of LIT

14

Abstract

Lymphocyte Immunization Therapy (LIT) was the recommended treatment for unexplained RPL in the nineties. After the REMIS study was published, the subsequent Cochrane reviews showed no benefit of LIT. However, there were many studies published afterward that showed significant benefits of LIT in RPL/RIF. In this chapter, a review is done of all the studies that were published in the last 20 years (Post-REMIS studies). This chapter also includes a special mention of Prof. Carp's important study and the study by Tanaka, which were published prior to the REMIS study.

Keywords

LIT studies · Post-REMIS studies · Review of LIT publications

Author	Publication	Year	Role of LIT
Carp HJA, Toder V, Torchinsky A et al.	Hum Reprod Vol.12, no.2, 250–255, 1997	1997	Beneficial-RPL
Tanaka T et al.	Osaka City Med J	1997	Beneficial-RPL
Ramhorst R et al.	Am J Reprod Immunol	2000	Beneficial-RPL
Hayakawa S et al.	Am J Reprod Immunol	2000	Beneficial-RPL
Clark DA et al.	HUMAN Reproduction Update. Vol.7, No.5	2001	Beneficial-RPL
Check JH et al.	Fertility and Sterility Vol.79, Suppl.2, April 2003	2003	Beneficial-RIF
Adachi H et al.	Clin Immunol.	2003	Beneficial-RPL (Secondary RPL)
Pandey M	International Pharmacology 4 (2004)	2004	Beneficial-RPL

(continued)

M. Raut, M. Raut, *Lymphocyte Immunization Therapy (LIT) in Reproductive Failures*, https://doi.org/10.1007/978-981-19-2960-1_14

Author	Publication	Year	Role of LIT
Takeshita T	Journal of Nippon Medical School, Oct 2004; 71 (5)	2004	Equivocal-RPL (more research required)
Christiansen OB et al.	Fertility and Sterility Vol 83, No.4	2005	Beneficial-RPL
Yokoo T et al.	Fertil Steril	2006	Beneficial-RPL
Gharesi-Fard B et al.	Iran. J. Immunol. Vol. 4 No. 3	2007	Beneficial-RPL
Nonaka T et al.	Am J Reprod Immunol	2007	Beneficial-RPL
Kling C et al.	Giornale Italiano di Obstetricia e Ginecologia	2014	Beneficial-RIF
Wu et al. (2014)	Reproductive Biology and Endocrinology 2014, 12	2014	Beneficial-RPL
Motak-Pochrzęst H et al.	Neuroendocrinol Lett	2015	Beneficial-RPL
Liang X et al.	Eur J Contracept Reprod Health Care	2015	Beneficial-RPL
Motak-Pochrzęst H et al.	Neuroendocrinology Letters	2015	Beneficial-RPL/RIF
Hua Y et al.	Int J Clin Exp Med 2016;9(6)	2016	Beneficial-RPL
Chen J-L et al.	International Immunopharmacology	2016	Beneficial-RPL
Cavalcante MB et al.	Archives of Gynaecology and Obstetrics	2016	Beneficial-RPL
MK Raut, MM Raut	Am J Reprod Immunol, 78:45–73	2017	Beneficial-RPL
Liu et al.	Reproductive Biology and Endocrinology (2017)15	2017	Beneficial-RPL
Gunther V et al.	Geburtsh Frauenheilk 2018;78	2018	Beneficial-RPL/RIF
Achilli C, Seshadri S et al.	Fertility and Sterility Vol110. No.6, November 2018	2018	Not beneficial-RPL/RIF
Marcelo Borges Cavalcante	Predictors of Therapeutic Success (AJRI 2018)	2018	Beneficial-RPL
MM Raut, MK Raut	ASRI-CSI2018	2018	Beneficial-RIF
M. Sarno	Eur. J. Obstet Gynecol. Reprod. Biol (2019)	2019	Beneficial-RPL
Raut MM, Raut MK	Am J reprod Immunol Vol 81, Suppl 1	2019	Beneficial-RPL
Saremi AT et al.	SJRM 2019, 3(1): 9–12	2019	Beneficial-RIF

14.1 Review of Studies Related to LIT

1. Carp HJA et al. (1997) [1]:
   Study
   In a study published in *Human Reproduction* in 1997, women with five or more miscarriages were studied. At the initial testing, they did not have anti-paternal complement-dependent antibody (APCA). They were given paternal

leukocyte immunization and effect on the live birth rate was studied in primary, secondary, and tertiary aborters.

Results

(a) Primary aborters showed a twofold increase in the live birth rate, with a 31% advantage.
(b) The live birth rate for Tertiary Aborters increased threefold with a 50% advantage.
(c) LIT was very effective in preventing miscarriage for primary or tertiary aborters who had at least five miscarriages [1].

2. Tanaka T et al. (1997) [2]

Study

This study, published in *Osaka City Med J.* 1997, included 72 patients with RPL who were treated with LIT.

Results

Pregnancy rate 87%

Live birth rate 65.27%

Conclusion: LIT is beneficial in RPL [2].

3. Ramhorst R. et al. (2000) [3]

Study

In this study, published in the *American Journal of Reproductive Immunology*, LIT was given to 92 women who had RSA (79 primary [PA] and 13 secondary aborters [SA]). The "control" group included 37 RSA couples.

Results

Pregnancy rate 58% in treated group v/s 46% in control group.

Live birth rate 88% in treated group v/s 71% in control group.

Conclusion

LIT beneficial in RPL [3].

4. Hayakawa S et al. (2000) [4]

Study

This study, published in the *American Journal of Reproductive Immunology*, 12 patients with RPL were treated with LIT.

Pregnancy rate 58.33%

Live birth rate 33.33%

Conclusion

LIT may play an important role in maintaining successful pregnancies by causing a shift from Th1- to Th2-dominant status [4].

5. Clark DA et al. (2001) [5]:

Study

This review included a critical analysis of current trials and their shortcomings, as well as an update to a meta-analysis. It also provided comments on possible new approaches.

Conclusion

There is level 1 evidence concerning Leukocyte Immunotherapy which appears to improve the chance of live birth if given to appropriate patients [5].

6. Check JH et al. (2003) [6]:

Study

Table 14.1 Double blind randomized trial of paternal lymphocyte immunotherapy for women with RSA

Different groups of women with RSA	Different therapies	Number of women with RSA in each group	Confirmed pregnancy	Live births	Abortion	Success (%)
1	AL	28	12	4	8	33
2	TPL	31	19	6	13	31
3	PL	32	25	21	4	84
4	NS	19	8	2	6	25
5	NT	14	9	4	5	44

AL autologous lymphocytes, *TPL* third party or unrelated male lymphocytes, *NS* sterile normal saline, *NT* not received any treatment or dropout patients, *PL* paternal lymphocyte, *RSA* recurrent spontaneous abortion, *MLR-Bf* mixed lymphocytes reaction blocking antibodies

In a matched controlled study published in *Fertility Sterility*, 37 women with previous two failed IVF-ET cycles were given LIT before the next ET.

Results

Clinical pregnancy rate: 70.3% v/s 45.9%

Live birth rate: 51.3% v/s 16.2%

Results showed very positive results after ET [6].

7. Adachi H et al. (2003) [7]

   Study

   In this study, published in *Clinical Immunology*, 26 patients with unexplained secondary RPL were treated with LIT. Of these, 22 patients conceived and delivered. The success rate was 90.9%. In the control group (11 patients), the success rate was only 18.2%.

   Conclusion

   Use of the husband's mononuclear leukocytes (LIT) appears to be an effective immunotherapy for unexplained secondary aborters [7].

8. Pandey MK et al. (2004) [8]:

   Study

   Published in the journal *International Immunopharmacology*, this study evaluated the effectiveness related to pregnancy outcomes, of paternal lymphocyte immunotherapy, for women suffering from Recurrent Spontaneous Abortions. This double-blind, randomized trial involved 124 women.

   Results

   The women who received LIT showed a significantly increased pregnancy rate (84%), compared to other groups, indicating the importance of LIT in women with RSA (see Table 14.1) [8].

9. Christiansen OB et al. (2005) [9]:

   Study

   This article, published in *Fertility and Sterility*, was the result of the workshop that took place in Denmark in 2002. It featured fertility specialists, obstetricians, and gynecologists from 18 countries.

   Outcome

There was some evidence that LIT using freshly prepared cells may be beneficial to subsets of primary RPL women. This is without evidence of anti-paternal cytotoxic immunity, as well as no autoimmune disease [9].

10. Yokoo T et al. (2006) [10]

    Study

    In this study, published in *Fertility and Sterility*, 52 patients with unexplained RPL were given LIT. Forty-two of them conceived and 34 delivered. Success rate was 80.76%. Eight patients miscarried again. Th1/Th2 ratio was significantly decreased in the successful group.

    Conclusion

    The husband's mononuclear cell (LIT) can be used to induce a dominant state in the patient's Th2 cells. This is beneficial for women with unexplained recurrent abortions [10].

11. Gharesi-Fard B et al. (2007) [11]

    Study

    The study included 92 nonpregnant patients with Recurrent Spontaneous abortions (three or more). They were given LIT with paternal lymphocytes. Eighty-one women with RSA were controlled (not immunized).

    Results

    Successful outcome was seen in 72.8% (67/92) of the immunized patients and 54.3% (44/81) in the control group, which was statistically significant ($p < 0.02$). Benefit was seen more in primary aborters than the secondary aborters.

    Conclusion

    The evidence in the study supports the use of LIT in improving the outcome of pregnancy in primary RPL patients [11].

12. Nonaka T et al. (2007) [12]:

    Study

    In this study, published in the *American Journal of Reproductive Immunology*, 165 patients with unexplained RPL and with negative MLR-Babs, underwent immunotherapy using freshly prepared paternal lymphocytes. One hundred forty patients conceived subsequently with successful pregnancy outcome in 110 cases (78.6%). The success rate was 30.0% in the control group.

    Conclusion

    In case of unexplained primary recurrent aborters, who do not have MLR-Babs, immunotherapy with paternal Lymphocytes (LIT) is very effective [12].

13. Kling et al. (2014) [13]

    Study

    This study, published in *Giornale Italiano di Ostetricia e Ginecologia* in 2014, is about the role of LIT in recurrent IVF implantation failures.

    It was a prospective observational experiment that involved 765 women who were nulliparous and went through LIT following 2.6 ± 0.8 stimulation cycles, and 4.1 ± 1.3% unsuccessful embryo transfers. Of all, 393 gave informative feedback after 2.1 ± 0.5 years.

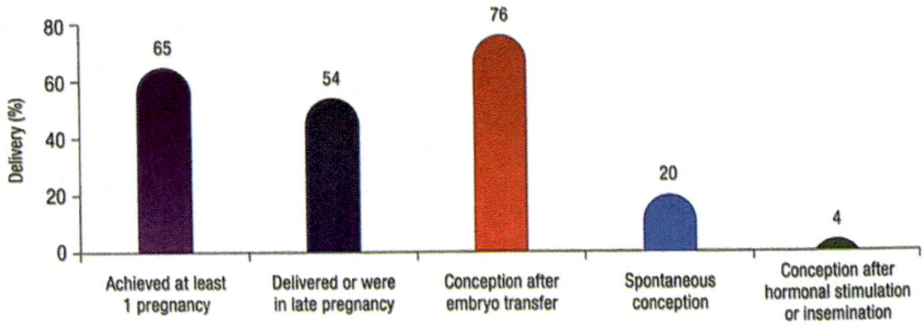

Fig. 14.1 Outcome 2 years after LIT in recurrent IVF failure

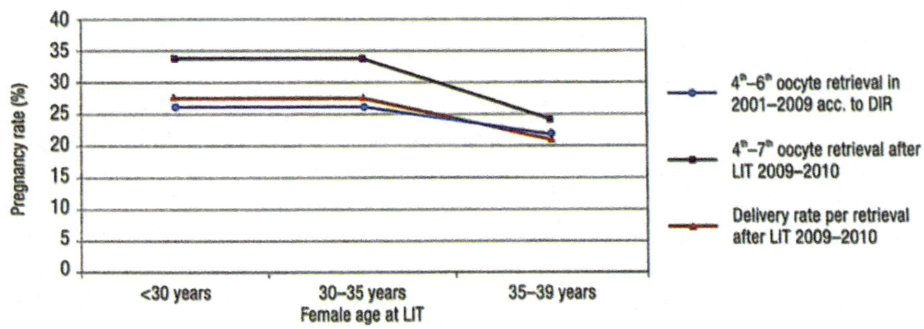

Fig. 14.2 Outcome after three unsuccessful OPU

Outcome: (Analyzed after 2 years) (see Fig. 14.1) [13]

(a) 65% women achieved at least one pregnancy
(b) 54% had delivered or were in late pregnancy (Fig. 14.1).
(c) Of all pregnancies, 76% were after ET, 20% were spontaneous, and 4% by ovulation induction by hormones or by donor insemination.
(d) In women up to 36 years, the success rate of LIT was elevated by a quarter in the first 6 months after LIT (see Fig. 14.2) [13].

14. Wu et al. (2014) [14]:

Study

The study, published in *Reproductive Biology and Endocrinology*, included 20 patients with unexplained RPL (3 or more losses and 7–12 weeks of pregnancy) and were given LIT with paternal lymphocytes. Analysis of Th17 and Treg cell populations was done by flow cytometry. Estimation of Th2 cytokine (TGF-β) and Th1 and Th17-associated cytokine (IL17A) was done.

Result

LIT led to a decrease in Th17 cells and an increase in Treg cells, leading to a decrease in Th17/Treg cell ratio, which is beneficial in the continuation of pregnancy. Also, Th1/Th2 cytokine ratio significantly decreased, protecting the pregnancy.

Conclusion

In unexplained RPL, LIT with paternal lymphocytes is beneficial by decreasing Th17/Treg cell ratio and reducing Th1/Th2 cytokine ratio, leading to immune tolerance [14].

15. Motak-Pochrzęst H et al. (2015) [15]

    Study

    In this study, published in the journal, *Neuroendocrinology Letters*, 241 women with RPL were given LIT. Of these, 202 women conceived and 169 had successful pregnancies (live birth rate 83.81%). One hundred fourteen patients from these 202 patients conceived for the second time later and 97 had a live birth (live birth rate of 80.7%).

    Conclusion

    LIT could improve the outcomes of pregnant patients with RM. It does not have any harmful effect on the immunological system of the patient. LIT also does not show any negative effects on the newborns, on long-term follow-up [15].

16. Liang X et al. (2015) [16]

    Study

    This study was published in *The European Journal of Contraception & Reproductive Health Care*. It involved 302 women who had suffered three or more losses in pregnancy. LIT was administered to three groups.

    First group was given lymphocytes donated by the partner, second group was given lymphocytes from a male donor, and third one from a woman donor. Fifty three patients were in the control group. The live birth rate was 87.7%, 75.8%, and 84.6%, respectively, in the three treated groups. The live birth rate was 40.3% in the control group.

    Conclusion

    Patients with spontaneous recurrent abortion can use female third-party lymphocytes for immunotherapy [16].

17. Motak-Pochrzęst H et al. (2015) [17]:

    Study

    In this study, published in the *Journal of Reproductive Immunology*, 100 patients with RPL and/or RIF were treated with LIT. There was a pregnancy rate of 44% and the live birth rate was 30%, indicating a beneficial effect of LIT [17].

18. Liu et al. (March 2016) [18]

    Study

    This study, published in the *American Journal of Reproductive Immunology* (2016), involved a meta-analysis of 18 randomized, placebo-controlled trials with LIT for Unexplained Recurrent Spontaneous Abortions (URSA). The live birth rates of each group were then calculated and then the Odds Ratio (OR) for LIT was found out.

    Results

Study or Subgroup	LIT Event	Total	Placebo Events	Total	Weight	Odds Ratio M.H. Fixed 95% CL	Year
Baozhu Liang 2006	42	50	26	50	8.3%	4.85 [1.90, 12.38]	2006
Xiaomei Liang 2011	52	64	26	64	9.7%	6.33 [2.84, 14.12]	2011
Lin Su2012	33	42	17	42	7.2%	5.39 [2.06, 14.09]	2012
Xiaoqin Liu 2012	51	57	10	42	2.4%	27.20 [9.01, 82.08]	2012
Yunxian Dai 2013	37	43	24	43	6.7%	4.88 [1.71, 13.98]	2013
Yunying Wang 2014	37	50	15	42	8.4%	5.12 [2.10, 12.51]	2014
Shuju Yuan 2014	31	38	15	38	5.5%	6.79 [2.38, 19.34]	2014
Zhiyang Jiang 2014	33	40	27	40	9.4%	2.27 [0.79, 6.49]	2014
Xiaofang Li 2014	46	53	28	53	7.4%	5.87 [2.24, 15.34]	2014
Hongbo Jing 2014	48	60	24	60	9.5%	6.00 [2.65, 13.58]	2014
Tianhong Wang2014	19	25	10	25	4.8%	4.75 [1.41, 16.05]	2014
Wei Cui2015	52	60	31	60	8.2%	6.08 [2.47, 14.96]	2015
Xiaoli DU 2015	32	40	16	40	6.4%	6.00 [2.21, 16.31]	2015
Shanshan Xu2015	18	25	11	25	6.1%	3.27 [1.01, 10.62]	2015
Total (95% CI)		**647**		**624**	**100%**	**5.75 [4.42, 7.40]**	
Total events	**531**		**280**				

Heterogeneity. Chi2 = 12.07, df = 13 (P = 0.52), I2 = 0%
Test for overall effect: Z = 13.29 (P < 0.00001)

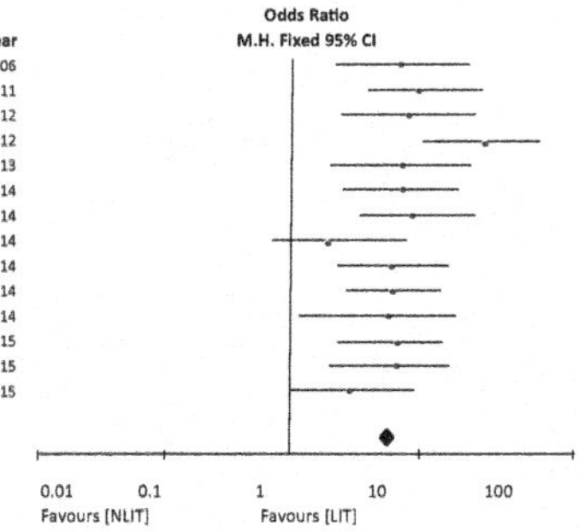

Fig. 14.3 Pregnancy rate in LIT (paternal and donor) v/s Non-LIT groups [19]

The treatment group had a significantly higher success rate (OR 3.74). LIT performed during and before pregnancy significantly increased the live birth rate (OR 4.67) [18].

(for details of the result, see Chap. 12)

19. Hua Y et al. (June 2016) [19]

Study

This study, published in the *International Journal of Clinical and Experimental Medicine*, involved a systematic review of studies to evaluate the efficacy of LIT in patients with Recurrent Spontaneous abortions without blocking antibodies (BA). It also included a meta-analysis comparing the efficacy of LIT and placebo in recurrent spontaneous abortions (RSA) in 14 double-blind randomized trials.

Outcome

A meta-analysis revealed that the LIT (paternal/donor) group has a higher pregnancy rate than the non-LIT group (OR 5.72 v/s 1.58 resp.) (see Fig. 14.3) [19]. The abortion rate was also found to be low in the LIT group (OR 0.19). Even in studies involving LIT with paternal lymphocytes only, the effect of LIT was beneficial (OR 5.09) (see Fig. 14.4) [19]. It also showed that the miscarriage rate was also significantly lower in the LIT group compared to non-LIT group (see Fig. 14.5) [19]. So, the meta-analysis concluded that LIT is a simple and economic method to treat recurrent abortions with negative blocking antibodies and it has good clinical application and good prospects for the patients [19].

20. Chen J-L et al. (2016) [20]

Study

The study, published in the journal, *International Immunopharmacology*, included 749 patients with unexplained RPL.

Of these, 380 were treated with LIT and 369 were controls (treated with progesterone). In the treated group, the pregnancy success rate was 89.7% while it was 32.2% for the control group. This suggests that LIT can effectively treat unexplained RPL.

Study or Subgroup	LIT Event	Total	Placebo Events	Total	Weight	Odds Ratio M.H. Fixed 95% CL	Year
Baozhu Liang 2006	42	50	26	50	9.5%	4.85 [1.90, 12.38]	2006
Xiaomei Liang 2011	52	64	26	64	11.1%	6.33 [2.84, 14.12]	2011
Lin Su2012	33	42	17	42	8.3%	5.39 [2.06, 14.09]	2012
Yunxian Dai 2013	37	43	24	43	7.6%	4.88 [1.71, 13.98]	2013
Shuju Yuan 2014	30	38	15	38	7.2%	5.75 [2.08, 15.88]	2014
Yunying Wang 2014	37	50	15	42	9.7%	5.12 [2.10, 12.51]	2014
Zhiyang Jiang 2014	33	40	27	40	10.8%	2.27 [0.79, 6.49]	2014
Hongbo Jing 2014	48	60	24	60	10.9%	6.00 [2.65, 13.58]	2014
Xiaofang Li 2014	46	53	28	53	8.4%	5.87 [2.24, 15.34]	2014
Shanshan Xu2015	18	25	11	25	7.0%	3.27 [1.01, 10.62]	2015
Wel Cui2015	52	60	31	60	9.4%	6.08 [2.47, 14.96]	2015
Total (95% CI)		**525**		**517**	**100%**	**5.09 [3.83, 6.76]**	
Total events	**428**		**244**				

Heterogeneity. Chi2 = 3.57, df = 10 (P = 0.52), I2 = 0%
Test for overall effect: Z = 11.25 (P < 0.00001)

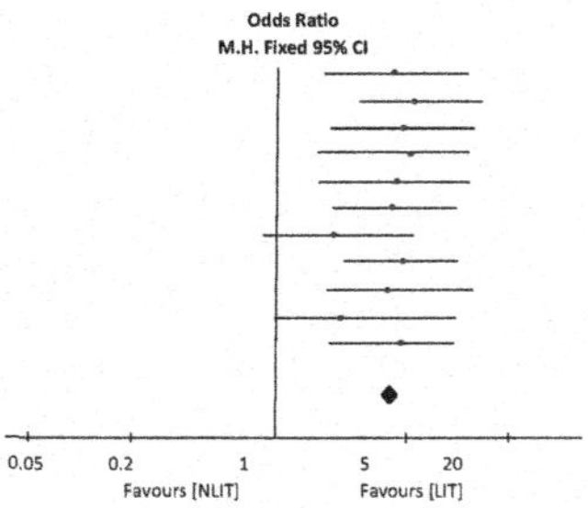

Fig. 14.4 Pregnancy rate in LIT (paternal only) v/s Non-LIT groups [19]

Study or Subgroup	LIT Event	Total	Placebo Events	Total	Weight	Odds Ratio M.H. Fixed 95% CL	Year
Baozhu Liang 2006	8	50	24	50	7.3%	0.21 [0.08, 0.53]	2006
Xiaomei Liang 2011	12	64	38	64	11.2%	0.16 [0.07, 0.35]	2011
Lin Su2012	9	42	25	42	7.1%	0.19 [0.07, 0.48]	2012
Xiaoqin Liu 2012	6	57	32	42	11.9%	0.04 [0.01, 0.11]	2012
Yunxian Dai 2013	6	43	19	43	5.9%	0.20 [0.07, 0.59]	2013
Shuju Yuan 2014	7	38	23	38	6.8%	0.15 [0.05, 0.42]	2014
Tianhong Wang2014	6	25	15	25	4.1%	0.21 [0.06, 0.71]	2014
Xiaofang Li 2014	6	53	19	53	6.1%	0.23 [0.08, 0.63]	2014
Yunying Wang 2014	10	50	18	42	5.7%	0.33 [0.13, 0.84]	2014
Zhiyang Jiang 2014	7	40	13	40	3.9%	0.44 [0.15, 1.26]	2014
Hongbo Jing 2014	12	60	36	60	10.4%	0.17 [0.07, 0.38]	2014
xiaoli DU 2015	8	40	24	40	6.9%	0.17 [0.06, 0.45]	2015
Wel Cui2015	8	60	29	60	9.1%	0.16 [0.07, 0.40]	2015
Shanshan Xu2015	7	25	14	25	3.6%	0.31 [0.09, 0.99]	2015
Total (95% CI)		**647**		**624**	**100%**	**0.19 [0.14, 0.24]**	
Total events	**112**		**329**				

Heterogeneity. Chi2 = 13.90, df = 13 (P = 0.38), I2 = 6%
Test for overall effect: Z = 12.76 (P < 0.00001)

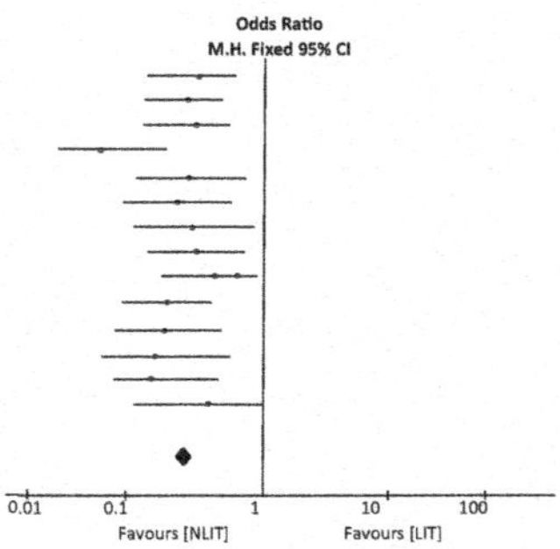

Fig. 14.5 Abortion rate in LIT v/s Non-LIT groups [19]

Conclusion

LIT is a treatment that effectively treats RSA and has a great clinical effect, according to the study [20].

21. Cavalcante MB et al. (December 2016) [21]

Study

This study, published in *Archives of Gynaecology and Obstetrics* (Dec 2016), is a systematic review of six published meta-analyses that included studies related to efficacy of LIT in Recurrent Miscarriages (RM). It analyzed the data related to efficacy and safety of LIT and all the statistical analyses were redone.

Results

(a) Two meta-analyses showed no improvement in the rate of live births in RM.
   (i) Fraser (1993): The effect of infusion of trophoblast Membrane or LIT was evaluated in four randomized trials. The results showed no improvement in live births.
   (ii) Wong (2014): This was the Cochrane meta-analysis, updated in 2014. It included 12 studies related to the role of LIT in RM. This meta-analysis showed no improvement in live birth rate with the use of LIT with unrelated donors in RM. (for details, see Chap. 12). One of the studies included in this meta-analysis was that of Christiansen et al.

performed with unrelated donor lymphocytes. However, the data that was shown from the study of Christiansen et al. was different from the one presented in the original article. This has impacted the outcome of the Cochrane meta-analysis adversely.

(b) Four meta-analyses found a beneficial effect of LIT in RM, with significant improvement in live births.
   (i) Collins J (1994)
   (ii) Coulam (1994): It is also known as RMITG meta-analysis (see Chap. 6)
   (iii) Pandey MK (2014): (See Chap. 13)
   (iv) Liu Z (2016): (See Chap. 13)

(c) New meta-analysis (Cavalcante MB): (See Chap. 12)

(d) Conclusion
   (i) The study concluded that LIT is a safe treatment.
   (ii) It further stated that LIT is a valid treatment for couples with RM, when performed with lymphocytes from fresh, non-stored blood.
   (iii) It is beneficial when given before and during pregnancy.
   (iv) The best results are seen when LIT is done with an adequate number of lymphocytes.
   (v) LIT is the most effective when applied intradermally.
   (vi) LIT is beneficial in well-selected couples (subgroups) [21].

22. Raut MK and Raut MM (Aug 2017) [22]:

    Study

    This study (abstract), published in the *American Journal of Reproductive Immunology* (Aug 2017), assesses the effect of LIT in RM. Couples with RM with negative lymphocyte crossmatch and/or raised NK cells and/or raised Sr. TNF-alpha and/or presence of endometrial CD57, were selected for LIT.

    Result: A success rate of 82% was seen in the study [22].

23. Liu et al. (2017) [23]:

    Study

    This study was published in *Reproductive Science and Endocrinology* (2017). The results evaluate the efficacy and safety of immunotherapy using a small dose of lymphocyte in patients who have unexplained recurring miscarriages (uRM), and Th1/Th2/Treg paradigm disorders. The outcome of 64 patients with uRM, given low dose LIT was compared with that of 35 patients, who did not receive LIT. Changes in Th1, Th2, and Treg cells after LIT were also noted.

    Outcome

    After LIT, the Th1 cell proportions decreased significantly while Treg and Th2 cells proportions increased significantly. The miscarriage rate was significantly less in the LIT group (11.6%), in contrast to 34.8% in the non-treated group ($p < 0.05$) (see Table 14.2) [23]. The study concluded that low-dose LIT is beneficial in restoring Th1/Th2/Treg paradigm and it improves the pregnancy outcome in patients with uRM [23].

Table 14.2 The proportion of Th1 cells, Th2 cells, and Treg cells before and after lymphocyte immunotherapy in patients with uRM [23]

	Before immunotherapy	After immunotherapy	*p* value
Total Th1	32.67 ± 9.05	31.38 ± 9.18	0.075
Total Th2	0.89 ± 0.40	1.00 ± 0.46	0.142
Total Treg	1.78 ± 0.64	2.33 ± 0.64	<0.001
Abnormal Th1 ($n = 33$)	37.69 ± 5.33	35.83 ± 6.86	0.038
Abnormal Th2 ($n = 14$)	0.44 ± 0.25	0.84 ± 0.42	0.013
Abnormal Treg ($n = 39$)	1.53 ± 0.30	227 ± 0.63	<0.001

24. Gunther V et al. (Jan 2018) [24]

    Study

    This study, published in *Geburtsh Frauenheilk* (2018), involved an extensive review and comparison of three meta-analyses—Wong (2014), Liu (2016), Cavalcante (2016), with the respective included studies.

    Outcome

    (a) LIT is a treatment option for unexplained RPL or RIF, though a significant benefit has not been proven due to heterogeneity of the data.
    (b) LIT may be effective if fresh lymphocytes are used.
    (c) LIT represents a safe and low-risk procedure.
    (d) Selection of couples for LIT can be done on a case-by-case basis, after thorough counselling of the couple regarding the chances of success [24].

25. Cavalcante MB et al. (Feb 2018) [25]

    Study

    This study, published in the *American Journal of Reproductive Immunology* (Feb 2018), is a retrospective, multicentric, observational study of 702 women with RM treated with LIT.

    Outcome

    The success rate of LIT was 60%. Four hundred twenty-one women successfully carried pregnancy till term.

    The factors predicting the success of LIT were as follows:

    (a) Maternal age:
        Higher the age, more the chance of LIT failure.
    (b) Number of previous miscarriages:
        In primary aborters, more the number of miscarriages, more is the chance of LIT failure.
        In secondary aborters, more the number of miscarriages (especially five or more), more is the chance of LIT success.
    (c) Associated autoimmune and thrombophilic factors reduce the success rate of LIT [25].

26. Raut MK and Raut MM (June 2018) [26]

    Study

    This study (abstract), published in the *American Journal of Reproductive Immunology* (June 2018), assesses the role of LIT in URIF. Sixty-five patients with URIF were given LIT, after selecting them with prior immunological testing.

Outcome

(a) Forty-eight patients were pregnant within 1 year. Forty-four of them conceived after ART and four spontaneously. There was a clinical pregnancy rate of 73.8%.
(b) The live birth rate was 67.7% (with three miscarriages and one tubal ectopic pregnancy) [26].

27. Achilli C et al. (Nov 2018) [27]

Study

This study, published in *Fertility and Sterility*, involved a systematic review of 30 randomized controlled trials (RCTs), satisfying the selection criteria and a meta-analysis was carried out of these studies.

Outcome

There is no role for Immunotherapy (including LIT), in improving the live birth rate of women who are undergoing IVF, with or without a history of RIF. It is also not useful in preventing a miscarriage in the case of idiopathic RPL. Better-designed RCTs and better patient selection are required to ascertain the role of LIT and other immunomodulatory methods [27].

28. Sarno M et al. (2019) [28]

Study

This study (2019), published in the *European Journal of Obstetrics & Gynaecology* and *Reproductive Biology*, is a retrospective analysis of a multicentric, observational study, involving 1096 couples who had a history of RM (two or more miscarriages) and negative lymphocyte crossmatch. The gestational and perinatal outcomes were studied.

Study included couples with RM with Lymphocyte crossmatch negative. They were divided into four groups:

(a) Alloimmune
(b) Alloimmune + Thrombophilia
(c) Alloimmune + Auto-immune
(d) Alloimmune + Thrombophilia + Autoimmune

Each group was divided into two subgroups, one was given LIT and the other did not receive LIT. Associated conditions were also treated. Gestational and perinatal outcomes were compared in all the groups (see Fig. 14.6) [28].

Outcome

(a) The LIT group had a higher live birth rate (60.1%) than the non-LIT group (33.1%).
(b) Sub-analysis of the above-mentioned subgroups showed that the LIT group had a significantly higher success rate in all immune categories, except in patients with autoantibodies.
(c) LIT's success rate was more if LIT was given during and before pregnancy, as opposed to just before pregnancy [28].

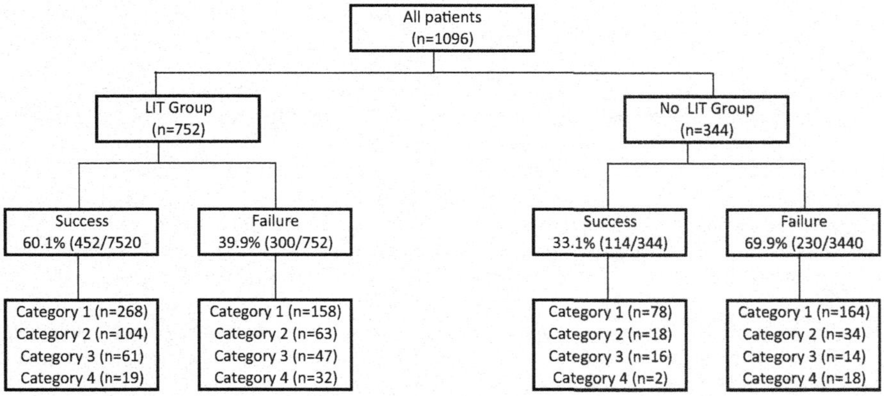

Fig. 14.6 Outcome in LIT and non-LIT groups and subgroups [28]

29. Raut MM and Raut MK (May 2019) [29]:
    Study
    This study (abstract—May 2019), published in the *American Journal of Reproductive Immunology*, assessed the efficacy of LIT in patients with URIF. Forty patients with URIF, who were chosen by immunological testing, received LIT before oocyte retrieval, if it was possible, or prior to frozen embryo transfer. Other immunomodulatory medications were used during the OPU cycle or ET cycle.
    Outcome
    The success of LIT was 80.7% in the patients, indicating that LIT can be effective in URIF patients, selected by prior immunological testing [29].
30. Saremi AT et al. (2019) [30]
    Study
    LIT was given to 61 patients with two or more IVF failures with negative anti-paternal cytotoxic antibodies. The mean of IVF failures was 2.95 ± 0.85.
    The pregnancy rate per LIT cycle was 31.1% (19/61), and the take-home baby rate was 84.2% (16/19). The overall full-term live birth rate per LIT cycle was 26.23% (16/61).
    Conclusion
    Patients who are experiencing unexplained recurrent IVF failures due to unknown reasons can have lymphocyte immunotherapy as a useful treatment option [30].

References

1. Carp HJ, Toder V, Torchinsky A, Portuguese S, Lipitz S, Gazit E, Mashiach S. Allogenic leukocyte immunization after five or more miscarriages. Recurrent Miscarriage Immunotherapy Trialists Group. Hum Reprod. 1997;12(2):250–5.

2. Tanaka T, Umesaki N, Maeda K, Miyama M, Ogita S. Pregnancy and neonatal outcomes in unexplained recurrent/habitual aborters treated by allogenic leukocyte immunization. Osaka City Med J. 1997;43:81–7.
3. Ramhorst R, Agriello E, Zittermann S, et al. Is the paternal mononuclear cells' immunization a successful treatment for recurrent spontaneous abortion? Am J Reprod Immunol. 2000;44:129–35.
4. Hayakawa S, Karasaki-Suzuki M, Ishii M, et al. Effects of paternal lymphocyte immunization on peripheral Th1/Th2 balance and TCR Vβ and Vγ repertoire usage of patients with recurrent spontaneous abortions. Am J Reprod Immunol. 2000;43:107–15.
5. Clark DA, et al. Unexplained sporadic and recurrent miscarriage in the new millennium: a critical analysis of immune mechanisms and treatments. Hum Reprod Update. 2001;7(5): 501–11.
6. Check JH, et al. Lymphocyte immunotherapy can improve pregnancy outcome following in vitro fertilization (IVF)-embryo transfer(ET)in patients failing to successfully conceive after two previous ETs. Fertil Steril. 2003;79(Suppl 2):12.
7. Adachi H, Takakuwa K, Mitsui T, et al. Results of immunotherapy for patients with unexplained secondary recurrent abortions. Clin Immunol. 2003;106:175–80.
8. Pandey MK, et al. Induction of MLR-Bf and protection of fetal loss: a current double blind randomized trial of paternal lymphocyte immunization for women with recurrent spontaneous abortion. Int Immunopharmacol. 2004;4:289–98.
9. Christiansen OB, et al. Evidence-based Investigations and Treatments of Recurrent pregnancy loss. Fertil Steril. 2005;83(4):821.
10. Yokoo T, Takakuwa K, Ooki I, et al. Alteration of TH1 and TH2 cells by intracellular cytokine detection in patients with unexplained recurrent abortion before and after immunotherapy with the husband's mononuclear cells. Fertil Steril. 2006;85:1452–8.
11. Gharesi-Fard B, et al. Effectiveness of leucocyte Immunotherapy on Primary recurrent Spontaneous abortion (RSA). Iran J Immunol. 2007;4(3):173.
12. Nonaka T, Takakuwa K, Ooki I, et al. Results of immunotherapy for patients with unexplained primary recurrent abortions—prospective non-randomized cohort study. Am J Reprod Immunol. 2007;58:530–6.
13. Kling C, et al. Lymphocyte immunotherapy (LIT) after recurrent IVF implantation failure: results of a prospective observational trial. Giorn It Ost Gin. 2014;XXXVI(6).
14. Wu L, et al. Alteration of Th17 and Treg cells in patients with unexplained recurrent spontaneous abortion before and after lymphocyte immunization therapy. Reprod Biol Endocrinol. 2014;12:74.
15. Motak-Pochrzęst H, Malinowski A. Polish experiences with paternal lymphocyte immunization in women with recurrent miscarriages. Neuroendocrinol Lett. 2015;36:572–7.
16. Liang X, Qiu T, Qiu L, et al. Female third party lymphocytes are effective for immunotherapy of patients with unexplained primary recurrent spontaneous abortion: a retrospective analysis of outcomes. Eur J Contracept Reprod Health Care. 2015;20:428–37.
17. Motak-Pochrzęst H, Malinowski A. Paternal lymphocyte immunization in women with recurrent embryo implantation failures and/or recurrent miscarriages after IVF. J Reprod Immunol. 2015;111:12.
18. Liu Z, Xu H, Kang X, Wang T, He L, Zhao A. Allogenic lymphocyte immunotherapy for unexplained recurrent spontaneous abortion: a meta-analysis. Am J Reprod Immunol. 2016;76: 443–53.
19. Hua Y, et al. Lymphocyte immunotherapy for recurrent spontaneous abortion in patients with negative blocking antibody. Int J Clin Exp Med. 2016;9(6):9856–67.
20. Chen J-L, Yang J-M, Huang Y-Z, Li Y. Clinical observation of lymphocyte active immunotherapy in 380 patients with unexplained recurrent spontaneous abortion. Int Immunopharmacol. 2016;40:347–50.

21. Cavalcante MB, et al. Lymphocyte immunotherapy in the treatment of recurrent miscarriage: systematic review and meta-analysis. Arch Gynecal Obstet. 2017;295(2):511–8. https://doi.org/10.1007/s00404-016-4270-z).
22. Raut MK, Raut MM. Efficacy of lymphocyte immunization therapy (LIT) in patients with recurrent miscarriages: Indian experience. Am J Reprod Immunol. 2017;78(1):62.
23. Liu M, et al. Low-dose lymphocyte immunotherapy rebalances the peripheral blood Th1/Th2/Treg paradigm in patients with unexplained recurrent miscarriage. Reprod Biol Endocrinol. 2017;15:95. https://creativecommons.org/licenses/by/4.0/
24. Gunther V, et al. Active immunization with partner lymphocytes in female patients who want to become pregnant—current status. Geburtsh Frauenheilk. 2018;78:260–73.
25. Cavalcante MB, et al. Lymphocyte immunotherapy for recurrent miscarriages: predictors of therapeutic success. Am J Reprod Immunol. 2018;79:e12833.
26. Raut MK, Raut MM. Role of immunomodulation treatment (Lymphocyte immunization therapy—LIT) in the treatment of unexplained recurrent implantation failure (URIF). Am J Reprod Immunol. 2018;80(S1):54.
27. Achilli C, et al. The role of immunotherapy in in vitro fertilization and recurrent pregnancy loss: a systematic review and meta-analysis. Fertil Steril. 2018;110(6):1089–100.
28. Sarno M, et al. Gestational and perinatal outcomes in recurrent miscarriages couples treated with lymphocyte immunotherapy. Eur J Obstet Gynecol Reprod Biol X. 2019;3:100036).
29. Raut MM, Raut MK. Immunomodulation with lymphocyte immunization therapy (LIT) and other immunomodulators in unexplained recurrent IVF failures—is it worth? Am J Reprod Immunol. 2019;81(Suppl 1)
30. Saremi AT, et al. Evaluation the success rate of treatment failure in patients with unexplained repeated IVF treatments immunization with lymphocytes (LIT) methods. SJRM. 2019;3(1): 9–12.

Case Studies 15

Abstract

This chapter includes two case studies depicting the efficacy of Lymphocyte Immunization Therapy (LIT). In the first case, the couple with 15 unexplained miscarriages underwent LIT and subsequently had a successful pregnancy. In the second case, LIT was beneficial for a couple with recurrent IVF failures. The case studies have discussed diagnostic assessment, LIT procedure and follow up. There is also a detailed discussion about various aspects of LIT. These case studies show that LIT can be beneficial in couples with "unexplained" repeated miscarriages or recurrent IVF failures as these can be due to allo-immune rejection of pregnancy.

Keywords

Repeated miscarriages · Recurrent IVF failures · Immunotherapy · Allo-immunity · Lymphocyte immunization therapy · LIT · Unexplained miscarriages · Unexplained IVF failures · Recurrent implantation failure

15.1 Case Study 1: Immunomodulation Using Lymphocyte Immunization Therapy, (LIT), in a Couple Experiencing 15 Repeated Miscarriages

Mugdha Raut and Mohan Raut

15.1.1 Introduction

Repeated miscarriages can be traumatizing for patients and can lead to permanent damage. In order to label a case RPL, the acceptance criteria are the number of miscarriages.

M. Raut, M. Raut, *Lymphocyte Immunization Therapy (LIT) in Reproductive Failures*, https://doi.org/10.1007/978-981-19-2960-1_15

American Society of Reproductive Medicine (ASRM) considers two or more miscarriages as RPL. However, the European Society of Human Reproduction and Embryology (ESHRE) uses three or more miscarriages for the definition of RPL [1]. With ASRM definition, incidents of RPL are 2%, while with ESHRE definition RPL affects 0.5–1% of pregnant women. As the number of miscarriages increases, it becomes increasingly difficult to treat as all the known tests to find the etiology are normal.

This case report discusses a case with 15 unexplained repeated miscarriages that resulted after spontaneous conceptions as well as after IVF treatment. All the known factors causing recurrent miscarriages were within normal limits. All the established modalities of treatment were unable to give success. In such cases, allo-immunity plays an important role and alloimmune rejection of pregnancy can explain the underlying cause. The case in discussion is unique in this matter, as the patient had experienced 15 miscarriages with failure of the established line of management and succeeded after immunomodulation treatment with lymphocyte immunization therapy (LIT).

15.1.2 Patient Information

Thirty-six-year old female residing out of India presented with a history of 15 miscarriages. The couple was married for 10 years. All the miscarriages were confined to the first trimester of pregnancy. The first eight miscarriages were following spontaneous conceptions. The next seven conceptions were following in vitro fertilization treatment. There were no previous medical conditions, including thyroid disorder, diabetes mellitus, or autoimmune diseases. There was no history of addiction to smoking or alcohol. There was no history of consanguinity or family history of similar problems. The patient had received all of the standard forms of treatment.

She was given progesterone, a low dose of aspirin, low molecular weight heparin, and vitamin supplements in the past pregnancies without success.

15.1.3 Clinical Findings

Clinical examination of the patient was normal.

15.1.4 Diagnostic Assessment

The couple had been investigated to rule out known factors of repeated miscarriages. The karyotype of both the partners was normal. Anatomical assessment of the uterus by ultrasonography and hysteroscopy was within normal limits. The tests for infections were normal. The thrombophilia profile did not show any abnormality. Tests to rule out autoimmune disorders did not reveal any defect. The couple was investigated at our center for alloimmune incompatibility. The lymphocyte crossmatch test was negative [1+]. The peripheral blood natural killer cells were tested. CD3 was 78%. Serum TNF-alpha was within normal limits. The infection profile (HIV, HBsAg, HCV, VDRL) of both the partners was normal. Blood group of the husband was A-positive and the wife was also A-positive. In view of the clinical history, negative lymphocyte crossmatch and a higher CD3 percentage, the couple was selected for LIT.

15.1.5 Therapeutic Information

The couple was given LIT at our center (ICPRM). The couple underwent LIT with husband's lymphocytes. The paternal lymphocytes were separated by the process of centrifugation using special media. The separated lymphocytes were repeatedly washed to remove the paternal plasma and RBCs. The paternal lymphocyte suspension is injected into the wife by intradermal route. The entire procedure of separation of the lymphocytes and injection of lymphocytes into the wife lasted for 4 h. There was only one sitting of the procedure.

15.1.6 Follow-Up

After the procedure was completed, the couple was observed to see if they had any reactions, but there were none. The couple was asked to follow up after 7 days to judge the response, especially for the intradermal injections in the form of erythema and induration. No medications were given to the patient after the procedure. The patient was asked to report any adverse reactions in the form of local and generalized pruritus, malaise, fever, and breathlessness. The patient reported none of these. No tests were repeated after the procedure. The couple was asked to make attempts at conception after a period of 4 weeks from the time of therapy.

15.1.7 Outcome

The couple went back to their country and planned pregnancy. The patient conceived spontaneously after 2 months [previous seven miscarriages were after IVF treatment]. The patient was advised to take progesterone, low dose of aspirin, low molecular weight heparin as soon as pregnancy was confirmed. Patient was on vitamins including folic acid. No booster dose of LIT was given. The pregnancy was followed up in her country. The pregnancy was uneventful. At 31 weeks, the fetus showed evidence of early IUGR for which she was given treatment. She delivered a baby boy at 37th weeks by LSCS. The operation was uneventful. The child is 5 years old now, is healthy and with normal milestones.

15.1.8 Discussion

The problem of repeated miscarriages is always a challenge to the treating physician. This area of reproductive medicine is the most challenging and frustrating.

It is difficult because the cause of the problem is not always known and there are no evidence-based diagnostic or therapeutic options. Studies on the etiology and management of RPL often fail to take into account all aspects [2]. Hence, the treatment of Recurrent Pregnancy Loss (RPL) can be frustrating many times. The major challenge is the diagnosis of the etiological factors as all the established ones are within normal limits. In such cases, the possibility of an alloimmune problem is very high. It has been shown that, as the number of miscarriages increases, it is more likely that the chance of immunological problems is there. In such cases, immunomodulatory treatment can play a very important role [3]. Of all the immunomodulatory treatments available, LIT presents an active form of treatment that can help to overcome this problem. LIT is a simple treatment, effective in properly selected patients, devoid of major side effects.

(a) *Literature*

Billingham and colleagues were the first to describe lymphocyte immunization. They found that skin grafts between fraternal twins were accepted. However, in the case of non-twin siblings, the skin grafts were not accepted [4]. Inadequate maternal recognition of paternal alloantigens could cause deficient tolerance to the pregnancy. It was, therefore, thought that immunization with paternal mononuclear cells might increase maternal recognition of paternal alloantigen. This would then allow the patients with recurrent spontaneous abortions, to continue pregnancy till term [5]. In 1981, lymphocyte immunotherapy (LIT) was used to treat four URPL patients. This was done based on the "tolerance" of human kidney allograft. Three of them delivered normal babies, while one gave birth to a premature baby [6].

Mowbray led the first randomized trial of LIT in unexplained, recurrent spontaneous abortions (RSA). Seventeen out of 22 women who were given paternal cells experienced successful pregnancies, compared to 10 out of 27 women who were given their own cells [7]. A collaborative study and meta-analysis in 1994 confirmed that LIT is effective in women with spontaneous primary abortions without anti-paternal antibodies. However, it was not effective in women who were positive for anti-paternal antibodies [8].

Allogeneic lymphocyte transplantation (ALT) was found to significantly increase the chance of live birth (OR 1.94) in patients with primary RM but no anti-paternal antibody. This was seen in a meta-analysis of placebo-controlled trials [9]. A meta-analysis of women with five or more miscarriages who were given LIT was done. It showed that the chance of live birth significantly increased following LIT. The beneficial effect was mainly seen in primary aborters [3].

(b) *Ober Study*
The results of this study showed that immunization with paternal mononuclear cells did not improve the chances of having a successful pregnancy in women who have had recurrent miscarriages. Nearly 65% of the control patients who were pregnant became successful despite having a history of unexplained recurrent miscarriages. Immunized women who were pregnant had a higher rate of miscarriage [10].

(c) *Criticism of Ober Study*
Ober's study is the only published study that found a negative impact of immunotherapy with lymphocytes on live birth rates. Ober study did not take into account important factors like the concentration of paternal Lymphocytes, storage of cells for many hours at temperatures ranging from 1 to 6 °C, the time between collection of blood and administration of immunizations, presence or absence of autoimmune disorders, nonidentical routes for immunotherapy (intradermal/subcutaneous and intravenous), number of doses along with lymphocyte concentration [11].

(d) *Cochrane Review 2003, 2006, and 2014*
Immunotherapy for recurrent miscarriages, paternal cell immunization, third-party donors leukocytes, trophoblast membranes, and intravenous immunoglobulin does not provide a significant beneficial effect, compared to placebo, for improvement in live birth rate.

(e) *Criticism of Cochrane Review*
The meta-analysis in the Cochrane review itself has been criticized by many researchers due to various flaws associated with it. Many subsequent studies have pointed out the deficiencies in the Cochrane Meta-analysis (separate analysis of primary and secondary RM was not carried out). It did not make distinctions between different immunizing routes and doses nor did it consider the total number of miscarriages. Instead, they have shown that Immunotherapy can actually play a positive role in recurrent spontaneous abortions [12].

It was later shown in a meta-analysis that immunization using lymphocytes led to a significant increase in live births. The rate was 77.8% for the group of women treated, in contrast to 46.1% for the control group. This was seen when Ober's study was removed from the Cochrane meta-analysis [13].

There was a new meta-analysis that included the data of Ober et al. It also included the data by Stray-Pederson et al., which was previously excluded by Liu et al. It showed that the increase in live births among couples who received immunotherapy (LIT), was still significant [14].

15.1.9 Rationale

In the current case, the role of the alloimmune factor as the etiology appears the most certain. Here, we have ruled out all the known factors responsible for repeated miscarriages. The patient had already received various available modalities of treatment that are used for such patients but without any success.

The lymphocyte crossmatch was negative; CD3 percentage was on the higher side of normal. The patient selection for LIT was done according to the following criteria:

1. History of 15 miscarriages.
2. First trimester miscarriages.
3. All routine investigations were normal.
4. Lymphocyte crossmatch was negative.
5. CD3 percentage high.

Other criteria that are important are (not present in this case):

6. Raised TNF-alpha.
7. Presence of endometrial NK cells [CD57].

15.1.10 Primary Takeaway Lessons

A case with a higher number of repeated miscarriages is more likely to be due to alloimmune rejection of pregnancy. If all the known factors for repeated miscarriages are ruled out, immunomodulation can help these patients to prevent subsequent miscarriages. Active immunotherapy in the form of LIT is still an effective therapy in properly selected cases.

15.1.11 Patient Perspective

When the patient first approached, she was totally frustrated and demoralized by the problem. She had found out about LIT by studying the literature on the Internet. She was willing to undergo any treatment that will help her without compromising her and her baby's health.

She was willing to come to India just to undergo this treatment. She had found out about the treatment and already discussed it with patients who had taken this treatment and had successful pregnancies. Now the patient is willing to share her experience with other patients suffering from similar problems who can be helped by LIT.

15.1.12 Informed Consent

Informed consent of the patient has been taken before the case study is prepared.

Conflict of Interest
There are no conflicts of interest.

15.2 Case Study 2: Role of Immunomodulation with Lymphocyte Immunization Therapy (LIT) in a Couple with Five Unexplained IVF Failures and 1 Miscarriage (IVF Pregnancy)

Mugdha Raut and Mohan Raut

15.2.1 Introduction

The problem of recurrent IVF failures and repeated miscarriages is traumatic to a patient. The number of pregnancy losses is the accepted criterion in labelling a case as RPL. American Society of Reproductive Medicine (ASRM) considers two or more miscarriages as RPL. However, the European Society of Human Reproduction and Embryology (ESHRE), uses three or more miscarriages for the definition of RPL [15]. Recurrent Implantation Failure, as it is currently defined, refers to a woman less than 40 years of age, who fails to conceive after three or more ETs or transfer of four embryos of good quality.

This case report discusses a case with a history of having done at least 20 IUIs, six IVFs of which only one resulted in a pregnancy which she miscarried. All the known factors causing IVF failure and miscarriage were within normal limits. All the established modalities of treatment were unable to give success. In such cases, allo-immunity plays an important role and alloimmune rejection of pregnancy can explain the underlying cause. The case in discussion is unique in this matter, as the patient had experienced 20 failed IUIs, 5 IVF failures, and a miscarriage with failure of the established line of management and succeeded after immunomodulation treatment with Lymphocyte Immunization Therapy (LIT) [16].

15.2.2 Patient Information

Thirty-six-year-old female, residing in India, presented with a history of 5 IVF failures and a miscarriage. The couple was married for 8 years. There was no history of any medical illness including diabetes mellitus, thyroid disorder, or autoimmune disease. No history of smoking or alcohol addiction was there. There was no history of consanguinity and no family history of similar complaints. The patient had undergone all the established forms of treatment. Without success, she was administered progesterone, low doses of aspirin, low molecular weight heparin, and vitamin supplements.

15.2.3 Clinical Findings

Clinical examination of the patient was normal.

15.2.4 Diagnostic Assessment

The couple had been investigated to rule out known factors of recurrent implantation failure and miscarriage. The karyotype of both the partners was normal. Anatomical assessment of the uterus by ultrasonography and hysteroscopy was within normal limits. The tests for infections were normal. The thrombophilia profile did not show any abnormality. Tests to rule out autoimmune disorders did not reveal any defect. The couple was investigated at our center for alloimmune incompatibility. The lymphocyte crossmatch test was negative [1+]. The peripheral blood natural killer cells were tested. CD3 was high. Serum TNF-alpha was within normal limits. The infection profile (HIV, HBsAg, HCV, VDRL) of both the partners was normal. Blood group of both partners was checked. The couple was chosen for LIT because of their clinical history, negative lymphocyte crossmatch, and higher CD3 percentage.

15.2.5 Therapeutic Information

The couple was given LIT at our center (ICPRM). The couple underwent LIT with husband's lymphocytes. The paternal lymphocytes were separated by the process of centrifugation using special media. The paternal lymphocyte suspension is injected into the wife intrademally. The entire procedure of separation of the lymphocytes and injection of lymphocytes into the wife lasted for 4 h. There was only one sitting of the procedure.

15.2.6 Follow-Up

After the procedure, the couple was observed for any reactions, which were none. The couple was asked to follow up after 7 days to judge the response, especially for the intradermal injections in the form of erythema and induration. No medications were given to the patient after the procedure.

The patient was asked to report of any adverse reactions in the form of local and generalized pruritus, malaise, fever, and breathlessness. The patient reported none of these. No tests were repeated after the procedure. The couple was asked to make attempts at conception after a period of 4 weeks from the time of therapy.

15.2.7 Outcome

The patient was conceived by IVF (ICSI) after 2 months. The patient was advised to take progesterone, low dose of aspirin, low molecular weight heparin as soon as pregnancy was confirmed. Patient was on vitamins including folic acid. No booster dose of LIT was given. The pregnancy was followed up regularly. The pregnancy went up to 35 weeks. Patient was delivered through an emergency LSCS, due to oligohydramnios with fetal distress. The child is 4 years old now, is healthy and with normal milestones.

15.2.8 Discussion

The problem of recurrent implantation failure is always a challenge to the treating physician. It is one of the most frustrating and challenging areas in reproductive medicine, as the cause of it is not always known. Also, the diagnostic and therapeutic methods that exist are not always evidence based. Hence, the treatment of recurrent implantation failure (RIF) can be frustrating many times [17, 18]. The major challenge is the diagnosis of the etiological factors as all the established ones are within normal limits. In such cases, the possibility of an alloimmune problem is very high. The immune system plays an essential role in the placentation process, including normal implantation and maternal placental fetal crosstalk [19]. It has been shown that, as the number of IVF failures increases, it is more likely that the chance of an immunological problem is there. In such cases, immunomodulatory treatment can play a very important role [20, 21]. Of all the immunomodulatory treatments available, LIT presents an active form of treatment that can help to overcome this problem [22]. LIT is a simple treatment, effective in properly selected patients, devoid of major side effects [23].

15.2.9 Lymphocyte Immunization Therapy

LIT has been studied extensively with respect to repeated miscarriages, which have an alloimmune basis. Inadequate maternal recognition of paternal alloantigens could cause deficient tolerance to the pregnancy. It was therefore argued that immunization with paternal mononuclear cells might increase maternal recognition of paternal alloantigen and allow patients with recurrent implantation failure or abortions to have a healthy pregnancy [24].

In the first randomized trial of LIT for unexplained recurrent spontaneous abortion (RSA), it was seen that 17 of 22 women who were provided with paternal cells had successful pregnancies. This was in contrast to the 10 of 27 who were given their own cells [7]. A prospective collaborative study with meta-analysis was done in 1994 to confirm the effectiveness of LIT for women who have had spontaneous abortions without anti-paternal antibodies but not in those with anti-paternal antibodies [8].

A meta-analysis of women with five or more miscarriages, who were given LIT, was done. It showed that the chance of live birth significantly increased following LIT. The beneficial effect was mainly seen in primary aborters [3].

In a study to assess the efficacy of LIT in recurrent implantation failure (RIF), it was seen that the success rates for women aged 36 years and older have increased by 25% in the first 6 months after LIT [25]. In one more study to know the role of LIT in RIF, LIT was given to 61 patients with 2 or more IVF Failures with negative APCA.

The pregnancy rate per LIT cycle was 31.1% and full-term live births per LIT cycle was 26.23% [26].

15.2.10 Rationale

The current case suggests that the role of the alloimmune factor is the most probable etiology. All known causes for miscarriage or repeated failures in implantation have been ruled out. The patient had already received various available modalities of treatment that are used for such patients but without any success.

15.2.11 Primary Takeaway Lessons

A case with a higher number of IVF failures and repeated miscarriages is more likely to be due to alloimmune rejection of pregnancy. If all the known factors for implantation failure or repeated miscarriages are ruled out, immunomodulation can help these patients to prevent subsequent failure. Active immunotherapy in the form of LIT is still an effective therapy in properly selected cases.

15.2.12 Informed Consent

Informed consent of the patient has been taken before the case study is prepared.

References

1. American Society for Reproductive Medicine. Definitions of infertility and recurrent pregnancy loss. Fertil Steril. 2008;90(Suppl 5):S60.
2. Christiansen OB, Nybo Andersen AM, Bosch E, et al. Evidence-based investigations and treatments or recurrent pregnancy loss. Fertil Steril. 2005;83(4):821–39.
3. Carp HJA, Toder V, Torchinsky A, et al. Allogenic leucocyte immunization in women with five or more recurrent abortions. Hum Reprod. 1997;12:250–5.
4. Billingham RE, Brent L, Medawar PB. Actively acquired tolerance of foreign cells. 1953. J Immunol. 2010;184:5–8.
5. Beer AE, Semprini AE, Zhu XY, et al. Pregnancy outcome inhuman couples with recurrent spontaneous abortions: HLA antigen profile; HLA antigen sharing; Female serum MLR blocking factors; Paternal leucocyte immunization. Exp Clin Immunogenet. 1985;2:137–53.
6. Taylor C, Faulk WP. Prevention of recurrent abortion with leucocyte transfusions. Lancet. 1981;2:68–70.
7. Mowbray JF, Gibbins C, Lidden H, et al. Controlled trial of treatment of recurrent spontaneous abortion by immunization with paternal cells. Lancet. 1985;1(8435):941–3.
8. Recurrent Miscarriages Immunotherapy Trialists Group (RTIMG). Worldwide Collaborative Observational study and meta-analysis on allogenic leucocyte immunotherapy for recurrent spontaneous abortion. Am J Reprod Immunol. 1994;32:55–72.
9. Daya S, Gunby J. The effectiveness of allogenic leucocyte immunization in unexplained primary recurrent spontaneous abortions. Am J Reprod Immunol. 1994;32:294–302.

10. Ober C, Karrison T, Odem RB, Barnes RB, Branch DW. Stephenson MD Mononuclear-cell immunisation in prevention of recurrent miscarriages: a randomised trial. Lancet. 1999;354: 365–9.
11. Clark DA. Cell-surface CD200 may predict efficacy of paternal mononuclear leukocyte immunotherapy in treatment of human recurrent pregnancy loss. Am J Reprod Immunol. 2009;61:75–84.
12. Christiansen OB. www.comtecmed.com/COGI/COGI9/Uploads/assets/christiansen%20ob.pdf
13. Liu Z, Xu H, Kang X, Wang T, He L, Zhao A. Allogenic lymphocyte immunotherapy for unexplained recurrent spontaneous abortion: a meta-analysis. Am J Reprod Immunol. 2016;76: 443–53.
14. Cavalcante MB, et al. Lymphocyte immunotherapy in the treatment of recurrent miscarriage: systematic review and metaanalysis. Arch Gynecal Obstet. 2017;295(2):511–8.
15. Hachem HE, Crepaux V, May-Panloup P, et al. Recurrent pregnancy loss: current perspectives. Int J Women's Health. 2017;9:331–45.
16. Raut MK, Raut MM. Role of Immunomodulation with LIT in a couple with 15 unexplained repeated miscarriages. TOG. 2017;5:14–7.
17. Magdi Y, El-Damen A, Fathi AM, et al. Revising the management of recurrent implantation failure through freeze-all policy. Fertil Steril. 2017;108(1):72–7.
18. Sun H, Lee K, Park I, Kim S, Chi H, Kim J, Cho J. Is immunotherapy useful in recurrent implantation failure patients; Especially with high natural killer cells? Fertil Steril. 2016; https://www.fertstert.org/article/S0015-0282(16)62350-6/pdf
19. Kwak-Kim J, Kim JW, Gilman Sachs A. Immunology and pregnancy losses: HLA, autoantibodies and cellular immunity. Madame Curie Bioscience Database. 2000–2013. p. 1–15.
20. Egerup P, Lindschou J, Gluud C, et al. The effects of immunotherapy with intravenous immunoglobulins versus no intervention, placebo, or usual care in patients with recurrent miscarriages: a protocol for a systematic review with meta-analyses, trial sequential analyses, and individual patient data meta-analyses of randomised clinical trials. Syst Rev. 2014;3:89.
21. Maejima M, Fuji T, Yamashita T, et al. Immunotherapy before and during pregnancy improves pregnancy outcome in women who suffer from recurrent abortion and did not benefit from immunotherapy before pregnancy. Am J Reprod Immunol. 1998;39(1):12–5.
22. Check JH, Liss JR, Check ML, et al. Lymphocyte immunotherapy can improve pregnancy outcome following embryo transfer (ET) in patients failing to conceive after two previous ET. Clin Exp Obstet Gynecol. 2005;32(1):21–2.
23. Raut M, Raut M. Immunotherapy in recurrent miscarriages, implantation failures and unexplained infertility. Indian Pract. 2014;67(2):107–9.
24. Carp HJA. Recurrent Pregnancy loss: causes, controversies, and treatment. Boca Raton: Taylor and Francis; 2015.
25. Kling C, Kabelitz D, Hedderich J. Lymphocyte immunotherapy (LIT) after recurrent IVF implantation failure: results of a prospective observational trial. Glorn It Ost Gin. 2014;36(6): 513–8.
26. Saremi A, Golnava M, Pooladi A, et al. The success rate of lymphocyte immunization therapy (LIT) in patients with unexplained repeated IVF failure. SJM. 2018;3(1):1–2.

Future of LIT

16

Abstract

Lymphocytes Immunization Therapy (LIT) is a unique immunomodulatory treatment for "unexplained" reproductive failures. Numerous studies have shown the efficacy and safety of LIT. It holds great potential in the treatment of patients with reproductive failures, who have not responded to other modalities of treatment. It is essential to study the changes induced by LIT at the molecular level. It is important to identify the subgroups that will be benefited from LIT, by doing the relevant investigations. With an increasing understanding of the immune system and the effect of LIT on the immune system, LIT can play a very important role in reproductive failures.

Keywords

Research in LIT · Mechanism of action of LIT · Future prospects of LIT

Pregnancy loss is a condition that can cause severe stress to the physical, mental, social, and economic health. Nearly 50% of pregnant women do not know what caused their loss and this adds to the suffering.

Lymphocytes Immunization Therapy is a unique treatment used in the management of Repeated Miscarriages/Recurrent IVF Failures/Unexplained Infertility. This therapy was recommended in these conditions in view of the immunological incompatibility between the mother and fetus. This immunological difference between the mother and fetus is usually not detrimental to pregnancy. The maternal immune response toward pregnancy is always that of tolerance and not activation. However, in some unfortunate couples, the immune protection is breached and leads to immune rejection of pregnancy.

Numerous studies show that LIT has the ability to prevent implantation failure and pregnancy loss. Recent studies showed that LIT causes maternal immune system changes to help preserve the pregnancy. This includes enhancing the expressions of

M. Raut, M. Raut, *Lymphocyte Immunization Therapy (LIT) in Reproductive Failures*, https://doi.org/10.1007/978-981-19-2960-1_16

APCA, PIBF, MLR-Bf, and Ab2 as well reducing the Th1/Th2 ratio and changing the cytokine balance toward Th2 dominance [1].

Numerous studies related to LIT have given varying results. It is mainly due to widely varying protocols. LIT, when applied correctly to properly selected patient groups, can still hold a great potential as a successful and secure treatment approach for patients who have failed to respond to less aggressive treatment methods.

The immunomodulation induced by allogeneic lymphocytes has been successful in overcoming the immune rejection seen in some women, thus preventing RPL/RIF/UI. Numerous studies as mentioned in previous chapters have proved the efficacy of LIT. Various recent meta-analyses have underlined the success of LIT in these conditions.

The bigger challenge that we have faced with LIT is to find the underlying mechanism that leads to success of LIT.

The changes induced by LIT at the molecular and cellular levels need to be researched further. It is important to find out which component of allogeneic lymphocytes is important for immunomodulation. Research should be more focused on the chemical response to the lymphocytes and which cytokines predominantly increase after LIT. It is then possible to use that component or cytokine to produce the immunomodulatory effect rather than injecting lymphocytes themselves.

It is important to identify which subgroups will benefit from LIT. For patients to be eligible for this treatment, it is very important to strictly decide the inclusion and exclusion criteria for LIT. To select couples for LIT, it is necessary to perform more precise diagnostic tests. Crossmatching the serum of patients with the partner's lymphocytes for detecting anti-lymphocyte maternal antibodies, mixed Lymphocyte Culture, evaluation of interleukins, and valuation of soluble CD30 dose is recommended as potential laboratory markers to increase the rate of success for LIT [1].

LIT can revolutionize the treatment of Infertility/Recurrent pregnancy loss/recurrent IVF Failure in future and can become the first line of treatment in Unexplained cases and even before ART procedures. As we know more and more about the intricacies of the immune system and the immune basis of different diseases, immunomodulation by LIT can play an important role in the fields of medicine other than Reproductive Immunology.

The important points to consider to make progress in the field of LIT in RPL and RIF are as follows:

1. Diagnostic tests to select the subgroups of patients, who will be benefited from LIT.
2. A common treatment protocol should be agreed upon.
3. Well-designed, randomized controlled trials to be conducted.
4. A global registry of all LIT patients treated at various centers should be done.
5. More research is required to find out the mechanism of action of LIT at the molecular level, which may help to replace lymphocytes, with a chemical that will produce the immunomodulation.

Reference

1. Hajipour H, Nejabati HR, Latifi Z, Hamdi K, Bahrami-Asl Z, Fatta A, Nouri M. Lymphocytes immunotherapy for preserving pregnancy: mechanisms and challenges. Am J Reprod Immunol. 2018;80(3):e12853.

The manufacturer's authorised representative in the EU is Springer Nature Customer Service Centre GmbH, Europaplatz 3, 69115 Heidelberg, Germany. If you have any concerns regarding our products, please contact ProductSafety@springernature.com

Printed and bound by CPI Group (UK) Ltd, Croydon, CR0 4YY
30/09/2026
02225308-0008